AF589462

Marine Products Sector in India

THE AUTHORS

Dr. V. P. Ancy is an aspiring researcher in the field Industrial Economics. She was awarded Ph.D. in Economics, in the area of "Fishing Industry in Kerala: The Perils of International Quality Assurance Standards". She is working as an Assistant Professor in the Research Department of Economics at Maharaja's College (A Government Autonomous College), Ernakulum. Dr. Ancy has published more than 10 papers in various reputed journals. She has done a minor project in the topic of "Challenges of Quality Assurance Standards in Seafood Export Industry in Kerala" Sponsored by University Grant Commission, New Delhi. She has presented research papers in various National and International Conferences and Seminars. Currently, she is a member of Board of Studies and the IQAC Economics Department Coordinator in Maharaja's College, Ernakulam, Kerala.

Dr. K. V. Raju is an Associate Professor in the Department of Economics at Sacred Heart College, Thevara, Kochi. Dr. Raju has published more than 65 papers in various national and international journals of Repute. Prof. Raju's interests include Economics of Education, Industrial Economics, International Trade and Agricultural Economics. He has produced Nine Ph.D.'s in various research areas. He has undertaken projects of various prestigious organizations. Dr. Raju is also a guide of M. Phil Programme, School of Distance Education, Madurai Kamaraj University and Ph.D.Programmes of M G University, Kottayam and Bharathiar University, Coimbatore and M. Phil Programme of Indira Gandhi National Open University, New Delhi. He has been selected for the Hubert Humphrey Alumni Award of the University of Minnesota in 2001 and the prestigious K K Birla National Fellowship in Economics 1999-2000. Dr. Raju was the Head of the Department of Economics at the Mahatma Gandhi University Research Centre in Economics, Sacred Heart College, Thevara, during the period 2004-2014. Currently, he is the member secretary of the Academic Council of Sacred Heart College, Thevara and also a member of the PG Board of Studies in Economics of Mahatma Gandhi University, Kottayam, Maharaja's College, Ernakulam, and Christ College, Irinjalakkuda, in Kerala.

Marine Products Sector in India

— *Authors* —

V. P. Ancy
K. V. Raju

2018
Scholars World
A Division of
Astral International Pvt. Ltd.
New Delhi – 110 002

© 2018 AUTHORS

Publisher's Note:

Every possible effort has been made to ensure that the information contained in this book is accurate at the time of going to press, and the publisher and author cannot accept responsibility for any errors or omissions, however caused. No responsibility for loss or damage occasioned to any person acting, or refraining from action, as a result of the material in this publication can be accepted by the editor, the publisher or the author. The Publisher is not associated with any product or vendor mentioned in the book. The contents of this work are intended to further general scientific research, understanding and discussion only. Readers should consult with a specialist where appropriate.

Every effort has been made to trace the owners of copyright material used in this book, if any. The author and the publisher will be grateful for any omission brought to their notice for acknowledgement in the future editions of the book.

All Rights reserved under International Copyright Conventions. No part of this publication may be reproduced, stored in a retrieval system, or transmitted in any form or by any means, electronic, mechanical, photocopying, recording or otherwise without the prior written consent of the publisher and the copyright owner.

Cataloging in Publication Data--DK
Courtesy: D.K. Agencies (P) Ltd. <docinfo@dkagencies.com>

Ancy, V. P., author.
Marine products sector in India / authors, Ancy VP, KV Raju.
pages cm
Includes bibliographical references.
ISBN 9789387057715 (International Edition)

1. Seafood industry--India. 2. Fishery products--India--Marketing. I. Raju, K. V. (Associate professor of economics), author. II. Title.

LCC HD9466.I43K47 2018 | DDC 382.437095483 23

Published by : **Scholars World**
A Division of
Astral International Pvt. Ltd.
– ISO 9001:2015 Certified Company –
4736/23, Ansari Road, Darya Ganj
New Delhi-110 002
Ph. 011-4354 9197, 2327 8134
E-mail: info@astralint.com
Website: www.astralint.com

PREFACE

The Fishing Industry in India is contributing significantly to the agricultural export of the country and thereby helping poverty alleviation and generating employment to millions of people in the coastal area. In India over 14.5 million people depend on the fisheries sector for their livelihood, besides an equal number of people engaged in ancillary activities in fisheries and aquaculture. This is the sector on which large majority of the marginalized community depend for livelihood. Fisheries exports form a sizable share for foreign exchange earning of the country. Fish, which is considered as "rich food for poor people" has been recognized as a Health Food and more and more people in developed countries are moving to fish from red meat. Seafood is highly perishable in nature and it requires different infrastructure for handling, value-addition, processing and marketing. New Economic Policy of 1990's led to the massive adoption of scientific invention and technology intervention had a core impact on the fisheries structure in the country. The growth of export of marine fish products from developing countries has been regularly threatened by various non-tariff barriers from the developed countries. An effective technology transfer inventions and innovation would play a crucial role in confronting a number of supply side obstructions and numerous demand side opportunities. Kerala's shrimp exports have largely landed in USA and Japan. EU is the primary market for cuttlefish and squid exports. Therefore we should always be updated with the changes coming in the International Quality Standards, particularly with reference to market access and product diversification. There should be a centralized agency to monitor hygiene and health practices for both domestic and international market. In order to improve the productivity, profitability and stability of Kerala's seafood industry, there is need for the following revolutions, viz. Productivity revolution, Quality revolution, Value addition Revolution (ready to eat and ready to cook products etc) for both home and global trade. Seafood Export Processing units require financial and technical assistance to implement International Quality Standards. A concerted approach is needed to promote private and public cooperation in establishing an efficient quality infrastructure for improving the seafood export. Financial Constraints weaken the strategies of fisheries sector and restrict the growth of export from Kerala. Both public and private participation is essential for the creation of modern infrastructure and efficient processing facilities. This book examines the growth and contribution of fish production and to analyse the exports of Marine Fish products in the economy of India and Kerala. This book also addresses the Quality Assurance Standards followed in variouscountries and its impact on Marine fish exports. This book reveals the impact of Anti-dumping laws under the WTO framework and its impact on Marine exports.The present study made an in depth analysis of marine export processing sector in Kerala. We are greatly reminisced, Dr.P Indira Devi,Professor of Agricultural University, Thrissur who strongly recommended to publish our research work. This books' potentiality can be defined by the reader himself/herself. We dedicate this book to the research scholars, students, government department, policy makers and other agencies concerned with the marine export sector.

V. P. Ancy
K. V. Raju

CONTENTS

LIST OF ABBREVIATIONS

AA	Approved Arrangement.
AD	Antidumping Duty
APEDA	Agricultural & Processed Food Products Export Development Authority.
AHPND	Acute Hepatopancreatic Necrosis Disease.
AQA	Approved Quality Assurance
AQIS	Approved Quality Assurance system.
AQIS	Australian Quarantine and Inspection Service.
AQSIQ	Administration of Quality Supervision, Inspection and Quarantine.
BIN	Business Identification Number.
BPL	Below Poverty Line.
BRC	British Retail Consortium.
CAC	Codex Alimentarius Commission.
CCAMLR	Conference on the Conservation of Antarctic Marine Living Resources.
CCP	Critical Control Points.
CCSBT	Commission for the Conservation of Southern Blue Fin Tuna
CFC	Common Fund for Commodities.
CFIA	The Canadian Food Inspection Agency.
CFP	The Common Fisheries Policy.
CGE	Computable General Equilibrium.
CGMPs	Current Good Manufacturing Practices.
CI	Court of International Trade.
CIFT	Central Institute of Fisheries Technology
CIT	Court of International Trade.
CMFRI	Central Marine Fisheries Research Institute.
CNCA	Certification and Accreditation Administration of China.
CSO	Central Statistical Organisation.

CVD — Countervailing Duty
DDA — Doha Development Agenda.
DEPB — Duty Entitlement Pass Book.
DGAD — Directorate General of Anti dumping and Allied Duties.
DGFT — The Director General of Foreign Trade.
DOC — Department of Commerce.
DSB — Dispute Settlement Body.
EC — European Community.
ECM — Error Correction Model.
EDI — Electronic Data Interchange
EEZ — Exclusive Economic Zone.
EFSA — European Food Safety Authority.
EGM — Export General Manifest.
EHEDG — European Hygienic Equipment Design Group.
EIA — Export Inspection Agency.
EIC — The Export Inspection Council of India
EMS — Mortality Syndrome.
EPA — Economic Partnership Agreements.
ESCAP — Economic and Social Commission for Asia and the Pacific.
FAO — Food and Agriculture Organisation
TAP — Technical Assistance Programme
FAPs — Fishery and Aquaculture Products.
FDA — The Food and Drug Administration.
FFDA — Fish Farmers Development Agency.
FPA — Food Processing Accreditation.
FPAs — Fisheries Partnership Agreements.
FSMA — Food Safety Modernization Act.
FTP — Foreign Trade Policy.
GAP — Good Agricultural Practices.
GATT — General Agreement on Tariffs and Trade.
GDP — Gross domestic product.
GHP — Good Hygienic Practices.
GMP — Good Manufacturing Practices.
GSDP — Gross State Domestic Product
GSP — Generalized System of Preferences.
GTAP — Global Trade Analysis Project.
HACCP — Hazard Analysis and Critical Control Points.
HHS — Health and Human Services.
HPP — High Pressure Processing Foods.
IATTC — Inter-American Tropical Tuna Commission.
ICAR — Indian Council of Agricultural Research
ICCAT — International Commission for the Conservation of Atlantic Tunas.
ICRIER — Indian Council of Research on Inter at IHS national Economic Relations.

ICTSD	International Centre for Trade and Sustainable Development.
IDRC	International Development Research Centre.
IEQs	Individual Export Quotas.
IFAD	Fund for Agricultural Development.
IFPRI	The International Food Policy Research Institute.
IHHNV	Infectious Hypodermal and Hematopoietic Necrosis Virus.
IISS	India International Seafood show.
INFOSAN	The International Food Safety Authorities Network.
IOTC	Indian Ocean Tuna Commission.
IPQC	In Process Quality Control.
IQF	Individually Quick frozen.
ITA	International Trade Administration.
ITC	International Trade Commission.
IUU	Illegal Unreported and Unregulated Fishing.
LC	Letter of Credit.
MAFF	Ministry of Agriculture, Forestry and Fisheries.
MAP	Modified Atmosphere Packaging.
MFA	Multi-Fibre Arrangement.
MHLW	Ministry of Health, Labour, and Welfare.
MIPQC	Modified in Process Quality Control.
MOIQ	Market-Oriented innovative quality
MPEDA	The Marine Products Export Development Authority.
MRPLs	Minimum Required Performance Limits.
MSC	The Marine Stewardship Council.
MSY	Maximum Sustainable Yield.
MTS	Multilateral Trading System.
NAFIQAVED	The National Fisheries Quality Assurance and Veterinary Directorate.
NAMA	Non-Agricultural Product Market Access.
NCAP	National Centre for Agricultural Economics and Policy Research.
NCCP	National Codex Contact Point.
NFDB	National Fisheries Development Board.
NFPDB	National Fish Processing Development Board
NMFS	National Marine Fisheries Service.
NOAA	National Oceanic and Atmospheric Administration
NRCP	National Residue Control Plan.
NTB	Non-Tariff Barriers.
NTMs	Non-Tariff Measures.
OECD	Organistion for Economic Co-operation and Development.
PHT	Pre-Harvest Testing.
PRP	Pre-requisite Programme
PUFAs	Polyunsaturated Fatty Acid.
QCIA	Quality Control Inspection in Approved Units.
RASFF	Rapid Alert System for Food and Feed.

RTA's	Regional Trade Agreements.
SAFTA	South Asian Free Trade Area.
SAPTA	South Asian Preferential Trade Area.
SCM	Subsidies and Countervailing Measures.
SEAI	Seafood Exporters Association of India.
SHG's	Self-Help Groups.
SHIS	Status Holder Incentive Scheme.
SIFFS	South Indian Federation of Fishermen Societies.
SPIC	Seafood Park India.
SPS	Sanitary and Phytosanitary.
SQF	Safe Quality Food.
SSA	Southern Shrimp Alliance.
SSOP	Standard Sanitation Operational Procedures.
TBT	Technical Barriers to Trade.
TCMP	Tanzanian Coastal Management Partnership.
TCMP	Technical Co-operation Mission Programme
TED	Turtle Excluder Device.
TRIMS	Trade Related Investment Measures.
TRIPS	Trade Related Intellectual Property Rights.
TUSMP	Technology Upgradation Scheme for Marine Products.
UDFDA	United States Food and Drug Authority.
UNCTAD	United Nations Conference on Trade and Development.
UNDESA	The United Nations Department of Economic and Social Affairs.
USFDA	US The Food and Drug Administration.
USDOC	US Department of Commerce's.
VASEP	Vietnam Association of Seafood Exporters and Producers.
VERs	Voluntary export restraints.
WSSV	White Spot Syndrome Virus.
WTO	World Trade Organization.
YHS	Yellow Head Virus.

LIST OF TABLES

LIST OF FIGURES

1

INTRODUCTION

Fishing Industry is an imperative segment for ensuring food and nutritional security, sustainable development and for alleviation of poverty. It is the key sector for generating employment opportunities for the vast majority of the population. This sector is recognizing profound changes and challenges at National and Global level. A dangerous mix of the global economic slowdown combined with stubbornly high food prices in many countries lead to 1.3 billion people living in extreme poverty, and nearly 900 million chronically undernourished. An additional 1 billion suffer from hidden hunger due to lack of vitamins and minerals (FAO, 2012). According to the United Nations Department of Economic and Social Affairs, the world population is expected to grow from the present 6.8 billion people to about 9 billion by 2050(UN-DESA 2009).It estimated that from 2007 to 2050 the population growth in Africa and in Asia will be at 1 billion and 1.2 billion respectively. Fisheries and aquaculture are a vital source of food and protein for billions of people worldwide, and they support the livelihoods of more than one out of ten people (FAO, 2012).

Global Scenario

Globally, fish production from capture fisheries and aquaculture was over 167.2 million tonnes in 2014 as compared to 19 million tonnes in 1950 and to contribute significantly to food security and adequate nutrition for a global population expected to reach 9.7 billion by 2050. Global total capture production in 2014 was 93.4 million tones and aquaculture was 73.8 million tonnes(FAO, 2016).Fisheries sector directly or indirectly contributes 200 million jobs globally. In 2014, 84 percent of the global population engaged in the fisheries and aquaculture sector was in Asia, followed by Africa 10 percent and Latin America and the Caribbean 4 percent. Of the 18 million people engaged in fish farming, 94 percent were in Asia. Recent reports highlighted that the tremendous potential of the oceans and inland waters, contribute significantly to food security and adequate nutrition for a global population is expected to reach 9.7 billion by 2050. World per capita fish consumption increased from an average of 9.9 kg in the 1960s to 14.4 kg in the 1990s and 19.7 kg in 2013, with preliminary estimates for 2014 and 2015 pointing towards further growth beyond 20 kg due to population growth, rising incomes and urbanization, expansion of fish production and more efficient distribution channels (FAO, 2016). Moreover, fish continues to be one of the most-traded food commodities worldwide with more than half of fish exports by value originating in developing countries.

Globalization envisages chop and change target and ties the sustainability of small firms to the competitiveness of the industries. To succeed in global markets, value chain must be able to move production to the consumer in a more unique form than the value chains in competing countries. Need of the hour is to develop appropriate technologies, create required infrastructure and to evolve institutional arrangement for production, post-harvest and marketing of high value perishable commodities and their value added products. Pay more attention to reduce post-harvest losses-compress supply chain by linking producers and markets to promote processing of marine fish products in production catchments to add value before being marketed and to develop small scale processing refrigerated chambers and cold storages using conventional and non-conventional sources. More focus would be given to primary and secondary levels of value addition and processing. About 32 percent of world fish stock are estimated to be overexploited, depleted, or recovering and the need to be urgently rebuilt. Overall fisheries and aquaculture support the livelihoods of an estimated 540 million people, or eight percent of the world population. Fish products continue to be the most-traded of food commodities, worth a record $ 102 billion in 2008, up nine percent from 2007(SOFIA 2010).

Fish and fisheries products have become the most important foreign exchange earner among all agriculture products traded by developing countries. Developing countries became major net exporters of fish products in the late 1990s (IFPRI, 2003). The fishery net exports of developing countries showed a continuing rising trend in the last decades, growing from USD 9.6 billion in 1989 to USD 16.8 billion in 1999 to USD 25.5 billion in 2009. About 200 countries reported exports of fish and fishery products in 2012. Fishery exports reached a peak of US$129.8 billion in 2011, up 17 percent in 2010, but declined slightly to US$129.2 billion in 2012 following downward pressure on international prices of selected fish and fishery products(FAO, 2014). In 2014, fishery exports from developing countries were valued at US$80 billion, and their fishery net-export revenues (exports minus imports) reached US$42 billion.

According to the World Bank, five years after the global financial crisis, the world economy is showing signs of bouncing back in 2014, pulled along by a recovery in high income economies. Exports reached a new record of more than US$136 billion, up more than 5 percent on the previous year. Since 2002, China has been the largest exporter, reached a new record with export valued at US$ 19.6 billion and imports at US $ 8.0 billion. Norway the second major exporter, Vietnam became the third major exporter, overtaking Thailand, which has experienced a substantial decline in exports since 2013, mainly linked to reduced shrimp production due to disease problems. India rank seventh position in the export of fish and fishery products. Developed countries continue to dominate world imports of fish and fishery products though their shares have decreased in the recent years. Their share of world imports was 85 percent in 1992 and 73 percent in 2012. The United States of America and the leading importer Japan both highly depend on the imports, for fish consumption which is about 60 and 54 percent of their total fish supply respectively.

Challenges and Threats of Marine Sector

It is comprehended that fisheries sector would have to face several challenges and threats, along with the opportunities that are deriving from both supply and demand perspective. The outcomes of WTO negotiations under the Doha round, Hong Kong development (HK) round (December 2005) and the changing European Union (EU) regulations are likely to place new hurdles on the marine exports emerging from developing countries (Kamat Manasvi et al 2007). Marine fish processors and exporters face complex negotiations at the WTO-GATS level on tariffs and fishery subsidies, and bilateral and regional negotiations with the EU in the formulation of Economic Partnership Agreements (EPAs) and Fisheries Partnership Agreements (FPAs). In addition, they need to comply with increased food safety standards (SEAI, 2010). Being a highly sensitive item from the health and environment point of view, compliance costs of the seafood industry are bound to be quite high in relation to other durable export from developing countries. The key issue concerning NAMA is that while developing countries protect their markets through higher tariffs, the main mode of protection for the developed countries is through non-tariff measures, particularly through the use of technical barriers.

The growth of export of marine fish products from developing countries have been regularly threatened by various non-tariff barriers from the developed countries. Non-tariff barriers relates to hygiene and food safety, in particular sanitary, traceability and quality control aspects. One of the major challenges faced by exporters of fish and fishery products in developing countries is the progressively stricter food safety requirements in major industrialized countries (Henson Spencer et. al, 2004). Developed countries have established new requirements for fish imports, including labeling requirements, Hazard Analysis and Critical Control Points (HACCP) plans, to alleviate consumers concerns. Meeting these new requirements for documenting the safe handling, processing, and origin of fish products requires considerable experience, skill, and investment (Delgado L Christopher et.al 2003). In order to cope up with the International quality standards, the challenges faced by export processing firms are to develop innovative technologies and efficient management to raise productivity to meet the growing demand for fish and fishery products at the lowest cost.

On 29 September 2008, the Council of the European Union adopted EC No. 1005/2008 "establishing a Community system to prevent, deter and eliminate illegal, unreported and unregulated fishing" referred as the IUU Regulation. The

IUU Regulation of 1st January2010 is intended to regulate the highly complex multi-channel fisheries supply system of the European Community (EC). Under the IUU Regulation the importation of fishery products into the EC will be allowed only when the import is accompanied by a catch certificate, completed by the master of the fishing vessel and validated by the flag state of the vessel. USA and Japan markets are also moving towards establishing IUU control restrictions similar to the EC's IUU Regulations. Illegal, unreported and unregulated (IUU) fishing and related activities threaten efforts to secure long-term sustainable fisheries and promote healthier and more robust ecosystems. The international community continues to express its grave concern at the extent and effects of IUU fishing. Developing countries, often with limited technical capacity, bear the impact of this IUU fishing, which undermines their limited efforts to manage fisheries, denies them revenue and adversely affects their attempts to promote food security, eradicate poverty and achieve sustainable livelihoods. Beyond national boundaries, there is increasing need for international cooperation to improve global fisheries management of shared marine fish resources and to preserve the associated employment and other economic benefits of sustainable fisheries. Recognizing this, the European Union and United States of America, as leaders in the global fish trade, undertook (in 2011) to cooperate bilaterally to combat IUU fishing by keeping illegally caught fish out of the world market. Strengthening fisheries management capacity is fundamental in developing countries in order to facilitate sustainable fisheries and to reduce the impacts of IUU fishing (FAO, 2012).

Globalization has brought closer the markets and there is high demand for seafood items all over the world. Marine products form an important group of primary commodity exported from India accounting for about 4 percent of the total export earnings. One of the major challenges faced by developing country exports of fish and fishery products is stringent food safety requirement particularly in major markets such EU, USA, and Japan. The Uruguay Round Agreement on the Application of Sanitary and Phytosanary measures (SPS Agreements) and the Agreement on Technical Barriers to Trade (TBT) adopted by WTO members in 1995 have given a new direction to the international food trade to ensure that requirement such as quality, labeling, methods of analysis applied to internationally traded goods. Other regulations include Certification of Sustainability, discards, and juvenile control mechanisms, elimination of child labour in pre-production, production, pre-processing, processing sectors, subsidy reduction and Anti-Dumping restrictions, increase in fish for re-export and even channelizing in the domestic marketing system. The present study is an attempt to capture the development in the seafood export industry in the world trade scenario and the overall impact and performance of marine exports. Fishing industry is facing a lot of problems. Demand side issues are Quality Issues, International Standards and Regulations, Labeling and Certification Requirements, SPS, Codex Standards etc. Supply side issues are low levels of mechanization, low productivity, varying quality safety and hygiene, inadequate infrastructure, inadequate supply and quality of raw material, inadequate access to finance, marketing and Government legislation etc. Thus it will be useful to analyze the problems and development problems of seafood export industry in Kerala.

The future of fisheries export would be influenced by the consistent compliance with HACCP and SPS standards. The present study intends to provide insight into the challenges faced by the Marine products industry in our country. With countries, especially the importing countries increasing product standards in terms of food safety, quality, traceability, certification and ecolabelling, it is becoming difficult for the small-scale farmers to comply with the standards needed by both export and domestic markets. Raising market standards should not form a barrier or additional impediment for entry of products from small-scale farming. There is a move to introduce certification of aquaculture products before they are exported or marketed as is presently being done in case of capture fisheries by the Marine Stewardship Council(MSC). If this comes into force, it will have an impact on small-scale farmers who will not be able to bear the additional cost involved. If this sector is not to be marginalized, the small farmers need to be informed and trained to comply with international food safety standards. The government needs to come up with a comprehensive food safety and quality programme. Steps should be taken to devise appropriate institutional mechanisms to bring scattered small producers and processors under a network to enable them to participate in the emerging processing procedure to reap the benefits of expanding global fish trade.

Various trade barriers in the form of quality related issues, anti-dumping duties and bonding requirements etc were the results India's share in the Global Seafood Export trade accounted for about 2 percent only. Enhancing and expanding the production from the traditional and new production sources, striving for quality assurance along the production chain and achieving value addition are the major thrust areas identified. The major constraints for expansion of the industry are the availability of land and fresh water which is already becoming a constraint and hence it is necessary to judiciously undertake intensification without damage to the environment and bring in to production the marginal lands which are not suitable for agriculture. The deteriorating water quality has significantly affected the scope of marine fish exports.

During the financial year 2014-15, exports of marine products reached an all-time high of USD 5511.12 million. Marine product exports crossed all previous records in quantity, rupee value and USD terms. Exports aggregated to 10,51,243 MT valued at Rs. 33441.61 crores and USD 5511.12 million. India has excellent processing capability with a total of 412 processing units of which 216 units are approved for export to EU. The total installed capacity is 14539 MT per day. All these units are HACCP complied and regularly monitored for compliance to the laid down national and international standards by the competent agency i.e. the Export Inspection Council of India. We have in India 435 exclusive cold storages for marine products of which 27 are approved for storage for products for export to EU. The infrastructure available for processing in India is sufficient to cater the requirement of the importing countries. However, the infrastructure for handling the fish till it reaches the processing centers is inadequate.

FAO Mission to India during November 2008 has also insisted Govt. of India to take suitable steps to ensure proper hygiene in fishing vessels. MPEDA had requested the State Governments to suitably amend the MFR Act of the respective states, incorporating hygiene and sanitation standards in fishing vessels. The other maritime states are requested to initiate suitable action in this matter. Maintaining hygiene and sanitation in the fishing harbours and fishing vessels operating from the harbours is very important for ensuring quality of the fish harvested and landed at the fishing harbours. Though sufficient trainings are imparted to the fishermen on on-board handling maintenance of hygiene and sanitation in fishing vessels, there needs to be an enforcement mechanism to see that the fishing boats comply with these requirements. State government/Ports are the registering authority for fishing vessels as required under the Marine Fisheries Registration Act of the various states. However under the MFR Act the standards of hygiene and sanitation to be maintained in fishing vessels are not stipulated.

The present infrastructure facilities for berthing and landing are adequate to meet only 20 percent of the requirement in marine sector. In order to develop adequate infrastructure facilities, to begin with, fully integrated logistic harbours to cater to the need of the fishing industry are the essential and immediate requirement. Such infrastructure facilities include adequate fishing quay, handling, sorting, weighing, packing, icing, transporting facilities, storage of catch in case of delay in transportation, dry warehouses for storing items required for fishing boats and other fishing gadgets. Dry dock facilities for ship building; repairs and engineering services, bunkering and water supply ex-quay, and net mending yards are also equally important requirements. Adequate facilities provided in the fishing harbours ensure that the time spent in the port by fishing vessels are kept to a strict minimum enabling more fishing days.

Analytical Framework of Quality Issues

The Export Inspection Council of India (EIC) introduced a system for approval of the fish landing centers. According to this system, the Fisheries Department concerned will submit an application to the local office of the Export Inspection Agency (EIA). The EIA will inspect the harbor and recommend corrective actions wherever needed and give their approval once their recommendations are implemented. The harbour is then monitored by the EIA at regular intervals to note the progress made and to suggest further modifications, if any. The long-term use of Internet technology offers much wider opportunities for the development of more sophisticated services at a cost-effective price. The integration of IT systems offers comprehensive stock tracking along the supply chain. This adds value to the catch once it arrives at the end customer/supermarket. The information that accompanies the stock allows both quality and health and safety standards to be maintained (Ecclestone 2001). Traceability can be implemented only if all the links of the supply chain maintains the records on a one step up and one step down(i.e., one step backward and one step upward.) basis. In order to ensure that such records are maintained and are available to the competent authorities on demand, we need to have control over all the links of the supply chain, where the State Government machinery can play a vital role.

Currently the share of value added products, marine product exports are 8percent and 10percent in terms of quantity and value respectively. The MPEDA in its vision document has envisaged that at least 75percent seafood export to be in value added form comprising of ready-to-eat and ready-to-cook items by the year 2012. It also aims to upgrade the packaging facilities/systems for increased competency and compliance to international standards. With ever-changing and faster lifestyle, there is an increased demand for ready-to-cook and ready–to-eat seafood commodities in the international as well as domestic markets. There is a huge market demand in convenience product, snack food sector and also in ingredient industry. Present generation prefer value added marine products as they add convenience to end-users. With a view to promote value addition of seafood in the country, MPEDA has launched a new Technology Upgradation Scheme for Marine Products(TUSMP) aimed at promoting investment in infrastructure development for value addition. The seafood processing and marketing have become highly complex and competitive and exporters are trying to process more value added products to increase their profitability. Value can be added to fish and fishery products according to the requirement of different markets. These products range from live fish and shellfish to ready to serve convenience products. In general value added food products are raw or pre-processed commodities whose value has increased through the addition of ingredients or processes that make them more attractive to the buyer and or more readily usable by the consumer. It is a production/marketing strategy driven by customer needs and perceptions.

There are lots of products of fishery origin that find use in pharmaceutical and cosmetic industry. They are chitin, chitosan, omega-3 fatty acid, agar, align, isinglass, etc, to name a few. These products are of high demand and value in their respective industries and add to the unit value of the seafood exported from India. Besides innovative packaging, methods also add value to the seafood in retail segments, the popular packaging in retail market include tray, pouch, vacuum packing, retort pouches, Modified Atmosphere Packaging (MAP), sous-vide, cans etc. One of the major problems faced by the processing sector is the scarcity of labour in isolated rural communities. Difficulties in attracting local labour reflect the low pay, the seasonal or casual nature of employment and the poor work environment compared with office or supermarket jobs. The high turnover of labour and high levels of absenteeism experienced in some plants adds significantly to labour costs. As a result, firms are now turning increasingly to agency labour and the employment of unskilled and occasionally illegal immigrant workers.

Structural and Quality Revolution in the fisheries sector will impact the long term economic growth and contribute to the reduction in inequalities among the various fisheries clusters. Fishing industry is facing a lot of demand and supply side constraints. This study analyzed the problems and development problems of seafood export industry in Kerala. International Standards for food sanitation and quality control measures are highly capital intensive pave way for new era of Kerala's seafood export industry. The specific aims of this book is to identify the Quality Assurance Standards faced by suppliers of fish and fishery products in their major export markets, predominantly those relating to regulatory and customer requirements. And also examines the strategies of exporters to comply with quality standards and other market requirements. The supply chain in India is not strong enough to meet these rigorous standards. The seafood processing and marketing have become highly complex and competitive and exporters are trying to process more value added products to increase their profitability. The government with the help of private sector needs to come up with a comprehensive food safety and quality programme. The study uses the secondary as well as primary data sources. Gravity model is used to analyzed the market wise flow of marine product exports from India. Regression Analysis is used to estimates the international quality assurance standards, anti-dumping duty and other indicators of export trends to identify the impact of marine export of India. The study also employs non-parametric tests (Kruskal Wallis H Statistic test) to study and to rank the various problems faced by the surveyed unit. Enhancing and expanding the production from the traditional and new production sources, striving for quality assurance along the production chain and achieving value addition are the major thrust areas identified.

The data collections were confined to five districts only, i.e., Ernakulam, Alappuzha, Kollam, Thiruvananthapuram, and Kozhikode from Kerala State. It can be extended to other state for identifying the actual scenario of the seafood processing export industry. This is an area which can be taken up by researchers to make an assessment of the impact of quality standards on marine product export trade and analyse the impact of cost and benefits in the seafood industry. The hidden cost like human rights abuses of fish and fishery export industry has to be analysed for attaining the welfare aspect of humanity. The trade facilitation problem faced by the seafood exporters has to be examined and find solution to such problem. The researchers tries to analyse the marketing channel and price spread of fish to realize the pulse of demand for and supply of fishery products both in international and domestic markets. The present study has relied on quantitative and qualitative methodology of data collection. Fish production and Export trends data analysis primarily depend only upon Secondary data. Economic performance of the seafood export processing industry has been analysed, based on the primary data collection. The Survey Data is collected from 48 percent of the total export processing industry in Kerala. The survey data not collected from the cooperative unit which is one the emerging unit of the seafood. The scope of the present study is limited to explore the extent of marine export on international trade and its various dimensions. Especially it studies the link between quality standards and its impact on fish and fishery export trend and the study tries to make a preliminary step in this regard. The study used limited statistical tools for the analysis of data. Despite the limitations, present study is a meticulous attempt identifies the problems faced by the seafood export processing industry.

Theoretical Outline of the Book

The present study observed that the Prebisch-Singer hypothesis is true in the case of domestic as well as international trade in the fish and fishery export industry. Prebisch and Singer sought to use their theory to explain the performance gap between developing and industrialized countries. The theory states that the terms of trade between primary goods and manufactured products deteriorate over time. The main exporters of frozen fish and fishery products (primary) are developing countries and the major exporters of value added fish and fishery product are from developed countries. Majority of the developed countries import primary fish and fishery products from the developing countries that do not have the means to produce value added fishery products to export, so will lose out in the long-run, as their primary fishery products will become relatively cheaper than the value added products. It is based on the observation that the income elasticity of demand for value added products is greater than that for primary fishery products. As incomes raise demand for value added products increases more rapidly than demand for primary fishery products. Similarly in the domestic seafood export processing industry in which value added product exporting companies will survive and earns profit in the long run and the seafood export processing industry which export primary fish products will deteriorate overtime due to the modern life style and change in the customer preference to consume value added products than the other. The Prebisch-Singer hypothesis normally refers to the claim that the relative price of primary commodities in terms of manufactures shows a downward trend. The relative price of the primary fish products in terms of value added products shows a downward trend. Moreover, the fruit of technical progress had enriched the industrialized world than a country like India.

It also uses the endogenous growth framework to study the economic performance of marine fish export processing industries to economic growth. Neo-Schumpeterian growth models employ temporary monopoly profits, which motivate the discovery of new technology, as the mechanism to indigenize the impact of technological progress on growth. This branch of new growth models introduces conditions of imperfect competition at the micro-level of production, emphasizing the importance of temporary monopoly power as a motivating force for the intentional investment of resources by profit-seeking firms or entrepreneurs in the innovative process. In these models, growth depends on the incentives to invest in improving technology. It is also important to note that the appropriability of new knowledge is related to lead times over rival inventors and imitators rather than to effective patent protection.

The gravity model is used to examine the effect of quality standards, regulations and other NTMs on the exports of marine products from Kerala. Both primary and secondary data are used to examine the problems of marine product export industry of Kerala, cost and production functional framework, capacity utilization, effects of changing quality standards and the anti-dumping issues on shrimp export to USA. Business cyclical fluctuations are revealed in the seafood export processing industry. Recently, the world economy is moving towards downswing of business cycle, the US president Obama has remarked that the Great Recession (2008-2012) is the greatest economic crisis since the Great Depression. Theories of business cycles investigated and explore the impact of seafood export processing industry from Kerala.

Plan of this Book

The book is organized into five chapters. Chapter 1 traces the growth profile of marine fish production and exports from Kerala. Chapter 2 provides a detailed overview of the Quality Assurance Standards followed in various countries and explores the impact on Marine fish exports from Kerala. Chapter 3 highlights the detailed account of Anti-dumping laws under the WTO framework and investigates its impact on Marine exports. Chapter 4 probes into the analyses production and problem of fish processing industry/exports units in Kerala. Chapter 5 offers conclusions from the discussions.

Keywords: International Quality Standards, Fish Production, Marine Export Trend, Anti-dumping duty, Supply Chain and Value Chain Framework, Problems of Seafood Export Processing Industry.

2

MARINE FISH PRODUCTION AND EXPORT

Indian Scenario

In recent years, Fishing Industry has emerged as a dynamic sector and is being considered as a strategic sub-sector for promoting agricultural diversification. Growing urbanization, globalization, new lifestyles, trade liberalization and the emergence of new markets and rapidly changing social structures had a major impact on the fisheries structure in the country. The country now occupies the seventh position in total world marine capture fisheries production and the second place in inland fish production and is next to China. The fisheries sector plays an indispensable role in the economy of India contributing around 1 percent of the GDP. Fish production has increased at a higher rate, compared to food grains, milk, eggs and others, but the consumption per capita per annum is still at 9 kg with the fish eating population estimated at 56 percent in India. Per capita fish consumption in India is very low when compared with most of the countries in Asia. In view of the low per capita consumption of fish in India, it is necessary for the government and the concerned agencies to bring awareness among the population about the health benefits of fish consumption and organize the sector for the availability of fish under hygienic conditions and develop value added and easy to cook foods.

It is expected that the fish requirement by 2025 would be of the order of 16 million tonnes, of which atleast 12 million tonnes need to come from the inland sector and aquaculture is expected to provide over 10 million tonnes. (Handbook ICAR, 2011). In India as against a production of 6.4 million tonnes at the end of X plan, it is estimated that 10 million tonnes of fish food need to be produced by the end of XI Plan period (2012) to meet the growing demand. Demand for fish and fishery products are increasing considerably, both at domestic and export front. This has been caused due to the health concerns and the perception of fish as a healthy food with high levels of digestible protein, polyunsaturated fatty acid (PUFAs) and cholesterol lowering capabilities. The `projected demand for fish in the country by 2012 is 9.74 million tonnes (Fisheries Division, 2006, NCAP, 2006) including 5.9 million tonnes for domestic market. Considering the implications of WTO, there will be greater demand for improved and value-added fish and fisheries products. The projected supply of fish is 9.60 million tonnes by 2012 with major share of 5.34 million tonnes from inland aquaculture followed by 3.10 million tonnes from marine fisheries.

Fish production in the country has been showing an increasing trend and has reached a record level of 7 million tonnes in 2010-2011. This is due to the commendable progress in the inland fisheries sector during the last decade. With most of the wild stocks from the seas and rivers having been overexploited and depleted, the onus of bridging the gap between supply and demand falls on aquaculture and culture based capture fisheries, for which excellent opportunities exist in India for meeting the domestic demand as well as that of export. Diversification of production by introducing a new commercial species, adoption of new technologies like cage culture and introduction of processing units for value added products could add new dimensions to the sector. Proper post-harvest handling, reduction of post-harvest losses and hygienic primary processing are important to ensure quality and to prevent wastage. Simultaneously, cold chain development and hygienic marketing arrangements are to be made to ensure adequate returns to the fishers and farmers by making available good quality fish at affordable prices to the consumers.

Kerala Scenario

In Kerala, fishing Industry occupies an important position in its economy. Kerala occupies the 2nd position in marine fish production. The fish production in Kerala during 2012-13 was 6.8 lakh tonnes. Kerala's share in the national marine fish production is about 18.22 percent in value terms. The water resources of this state comprise of a coastline of 590 Km length having a continental shelf area of the sea adjoining the Kerala state is 39139 sq kms. Fisheries sector contribute 3 percent of the GSDP of the state. Kerala is the major producer of frozen shrimp in India, accounting for approximately 42 percent of national production. Kerala has an abundant scope to develop brackish water, fresh water and reservoir fisheries. However, there are non-utilisation and underutilisation of inland fish resources. According to the Kerala Marine Fisheries Regulation Act, the inshore area coming within the depth range of 50 meters has been demarcated for fishing by the traditional fishermen using country crafts and the area beyond the limit in the economic zones can be utilized by motorized boats and large vessels. As this restriction is not being strictly followed, monsoon trawling has been banned from 1980 onwards as a preventive measure. Although the fish catches from the Kerala coast include more than 300 different species, the commercially important number is about forty only (Economic Review, 2013). Table 2.1 gives the present fishery resources of Kerala and India and the percentage share from Kerala.

Table 2.1: Fishery Resources of Kerala and India

Marine	Kerala	India	Kerala's Percentage (Percent) Share
Length of Coast Line (Km)	590	8118	7.27
Continental Shelf('000 sq km)	40	530	7.54
Number of Fish Landing Centre's	187	1537	12.16
No of Fishing Villages	222	3432	6.47
No of Fishermen Families	118937	874749	13.59
Fisher-Folk Population	610165	4056213	15.04
Inland			
Total Inland water bodies(Lakh Ha)	5.43	73.59	7.38
River & Canals(Km)	3092	195210	1.58
Reservoirs(Lakh Ha)	0.3	29.07	1.03
Tanks & Ponds(Lakh Ha)	0.3	24.14	1.24
Flood Plain lakes/Derelict waters(Lakh Ha)	2.43	7.98	30.45
Brackish Water(Lakh Ha)	2.4	12.4	19.35

Source: Department of Animal Husbandry Dairying and Fisheries Handbook on Fisheries Statistics, Government of India 2014

With the declaration of the Exclusive Economic Zone (EEZ) in 1977, an area of 2.02 million sq km, comprising of 0.86 million sq. km on the west coast, 0.56 million sq.km on the east coast and 0.60 sq.km around the Andaman & Nicobar Islands has come under Indian jurisdiction. India is one of the coastal nations which witnessed rapid development of marine fisheries in the post-EEZ era. Government has also enacted the following legislation for the judicious exploration, exploitation, conservation and management of marine living resources. They are Marine Products Export Development Authority Act, 1972, The Wild life Protection Act, 1973 and various central legislation on environmental protection, Indian Coast Guard Act, 1978, The Maritime Zones of India (Regulation of Fishing by Foreign Fishing Vessels) Act, 1981, Marine Fishing Policy 2004, The Kerala Aquarian(Fisheries) Reform Act 2008 and so on. The coast of Kerala constitutes 7.27 percent of India's total coastline. Geographically, inland fisheries have great scope in the State. Kerala has 44 rivers, of which 41 originate from the Western Ghats and flow towards west into the marine flora and fauna of rich Arabian Sea. Table 2.2 gives the present state -wise potentials of the brackish water area available and the extent of its utilization. Aquaculture contributed more than 50 percent of the country's total fish production of 9.58 million tonnes during 2013–2014. The total aquaculture production of 4.43 million tonnes was valued at US$ 3.5 billion of which carp alone was responsible for as much as 4.18 million tonnes (FAO Statistics, 2014).

The coast of Kerala constitutes approximately 10 percent of India's total coastline. This coastline of 590 km and the Exclusive Economic Zone (EEZ) extends up to 200 nautical miles far beyond the continental shelf, which covers an area of 218536 sq. km. and provides opportunities in traditional fishing in inshore waters from ages. The continental shelf area is 39139 sq.km. The area within the 18m depth range accounts for 5000 sq.km, the area between 18-73m is approximately 25000 sq.km. and 73-182 m is the balance area. The Nine coastal districts of Kerala are Thiruvananthapuram, Kollam, Alappuzha, Ernakulam, Thrissur, Malappuram, Kozhikode, Kannur and Kasaragod. According to CMFRI census 2010, Kerala has 1,18,937 fishermen families, 222 marine fishing villages, and 187 marine fish landing centers.

Table 2.2: State-wise Estimated Potential Brackish Water Area.

	Estimated Area (ha)	Potential Area (Percent)	Area Developed (ha)	Area Developed (Percent)	Available Potential (Percent)
West Bengal	4,05,000	34.01	50,405	12.5	12.44
Orissa	31,600	2.65	13,400	40.8	42.4
Andhra Pradesh	1,50,000	12.6	84,951	51.12	56.63
Pondicherry	800	0.07	144	16.3	16
Tamil Nadu	56,000	4.7	6,104	9.4	10.9
Kerala	65,000	5.46	14,875	21.7	22.88
Karnataka	8,000	0.67	1,945	23.9	24.31
Goa	18,500	1.55	340	1.7	1.84
Maharashtra	80,000	6.72	1,135	1.6	1.42
Gujarat	3,76,000	31.57	2,371	0.6	0.63
Total	11,90,900	100	175,670	13.9	14.75

Source: MPEDA, Vision Document, Government of India, Kochi

Christians constituted 43 percent of fishermen families followed by Hindus 29 percent and 28 percent Muslims. There were 21,781 crafts in the fishery of which 4, 722 were mechanized, 11,175 motorized and non-motorized formed the rest. In Kerala, out of the 3,678 trawlers, 28 percent belongs to Ernakulam, 26 percent each to Kollam and Kozhikode districts which together constitute 80 percent. There were 13,985 crafts owned by fisher folk, of which 780 were mechanized, 6,671 motorized, and 6,534 non-motorized. There were 351 ice factories, 270 curing yards, 192 peeling sheds, 119 boatyards, 40 freezing plants and 28 cold storages in Kerala. Table 2.3 depicts fisher population in Kerala. While marine fish production in Kerala tended to fluctuate, the inland fish production showed a sign of improvement from 1999-2000.

Table 2.3: Fisher Population in Kerala

Population	Number
Fisher Population	6,10,165
Men	3,10,390
Women	2,99,775
Fishing Villages	222
Fishermen Families	1,18,937
Traditional Fishermen Families	1,16,321
Actual Fishing Full Time	1,30,922
Actual Fishing Part Time	1,582
Fish Seed Collection Full Time	2,991
Fish Seed Collection Part Time	901
Fish Landing Centers	187
Active Fishermen	1,45,396
Marketing of Fish	20,418
Repair of fishing nets	3,368
Curing/Processing	5,677
Peeling	9,817
Labourer	14,391
Others	736
Other than fishing	10,693
Religion-Hindu	34,509
Religion-Islam	33,708
Religion-Christianity	50,720
Caste-SC/ST	2,239

Source: CMFRI. Marine Census- 2010, Kochi.

During 2010-11, the marine fish production has decreased to 5.6 lakh tonnes from 5.70 lakh tonnes in 2009-10. Inland production sustained an increasing trend. District wise marine fish production showed that Alappuzha contributed the highest (23.08 percent) followed by Kollam (20.35 percent) and Kozhikode (14.95 percent). During 2010-11 the share of inland fish production to the total fish production of the state was 17.78 percent. The details of fish production for the last eight years are given Table 2.4 and Figure 2.1

Table 2.4: Fish Production in Kerala during the last Eight years

Year	Marine	Inland	Total
2007-2008	5.86	0.91	6.77
2008-2009	5.83	1.03	6.86
2009-2020	5.7	1.17	6.87
2010-2011	5.6	1.21	6.81
2011-2012	5.53	1.4	6.93
2012-2013	5.31	1.49	6.8
2013-2014	5.22	1.86	7.08
2014-2015	5.24	2.02	7.26

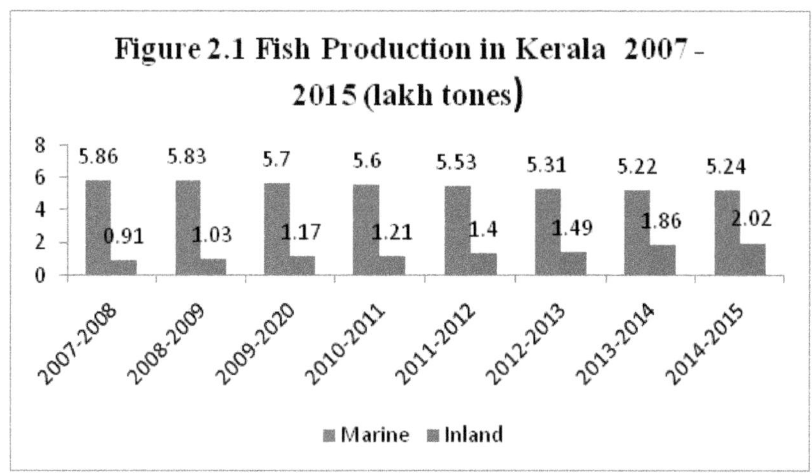

Source: Directorate of Fisheries, Government of Kerala, Thiruvananthapuram

Export Trends from India and Kerala

The marine fisheries sector in the country contributes about 81 percent of the total fish production and is one of the major contributors to foreign exchange earnings through seafood export. It constitutes about 16 percent of the total agriculture products export. Over 55 varieties of marine products are exported to different countries in South East Asia, Europe, China, Japan and USA. The total value of marine products export from the country was Rs.30213.26 crore in 2013-14 which accounts for approximately 1 percent of the total export from India. The quantity of marine exports from India during 2013-14 was 983756 tonnes. When compared to the previous year, it recorded a growth of 60.23 percent in rupee terms, 5.98 percent in volume and 42.60 percent growth in US$. During 2013-14, the export earnings have crossed 5 billion US dollars. Frozen shrimp is continued to be the single largest item of export in terms of value. It accounts 64.12 percent of the total U S $ export earnings. The figures must be viewed in the light of the scenario of continuing recession in the international markets, debt crisis in EU economies, continuing antidumping duty in US and the sluggish growth in US economy and the political instability in the Arab world. The increased production of Vannamei shrimp, increased productivity of Black tiger shrimp and better price realization of major items like Cuttlefish, Shrimp and Squid helped us to gain such a higher export turnover. The other major items of export are frozen fin fish, frozen cuttlefish, etc.

During 2002-03 and 2003-04 USA emerged as the single largest market for India's marine products and during 2004-05, the European union has collectively became the largest importer of Indian marine products and retained its position since 2005-06. During 2013-14, South East Asia continued as the largest market with a share of 26.38 percent in value followed by USA (25.68 percent), Japan (8.21 percent),Other Countries (8.20 percent), China (5.85 percent), and Middle East (5.45 percent).US had registered an incredible growth of 19.94 percent in quantity and 72.06 percent in US $ realization in frozen shrimp export. Export of Vannamei shrimp showed a remarkable increase in US market by 59.63 percent in quantity and 135.71 percent in US $ realization. There is a significant increase in exports to African countries in comparison to previous years, although the total exports to Africa remains very low as compared to other regions. Marine products were exported through 26 sea/air/land ports. Pipavav is the major port in terms of quantity (25.27 percent) and Vizagapatam is the major port in terms of dollar value (22.59 percent). Increased production of Vannamei shrimp, quality control measures and increase in infrastructure facilities for production of value added items have helped MPEDA to achieve the predicted target of USD 6 Billion for the year 2014-15.

The demand for seafood can be greatly enhanced by rigid quality control, hygiene and sanitation at the primary landing centers, harbours and processing plants. The quality of fish should be maintained at a very high level right from the capture by adopting various techniques. Bioterrorism, eco-labelling, traceability and risk assessment are some issues to be addressed. In the light of the HACCP regulations, the governments as well as industries are complying with the quality standards for the export products. Export market is subject to risk in terms of detention and loss or damage in transit or variation of exchange value. It is required to develop internal supply chain mechanisms with international specifications to promote sustainable internal marketing system. The consumers in developing countries are also becoming increasingly quality conscious due to spread of globalization and they are also aware of their rights and claim good quality produce for consumption (Sathiadhas and Femeena, 2002). They found that promotion of value added products not only accelerate forex earnings, but also provide a multiplier effect on employment front especially for weaker sections and women folk. Currently the export of byproduct constitute less than one percent of the total marine products exports. Considering the diminishing catch and over exploitation in the marine fisheries sector, the marine byproducts offer promising scope for ensuring alternate employment opportunities and income for the coastal fisher folk (Sathiadhas and Ashwathy, 2004).

The marine products export from the State during 2013-14 was 1.66 lakh tonnes, an increase of 18 percent over previous year. It is valued at 3435.85 crore constituting 17.93 percent in terms of volume and 18.22 percent in terms of value to Indian marine product export. The major export item is frozen shrimp during 2010-11. The state's share in all India exports has been declining in recent years. The share declined from 20 percent in quantity terms in 2000-01 to 17.93 percent in 2012-13 and the share in value increased to 18.52 percent from 16 percent. Exports of marine products from Kerala and India during 2010-11 are shown in Table 2.5.

Although the fish catches from the Kerala coast include more than 300 different species, the commercially important number is about forty only (Economic Review, 2013). Kerala's share in the marine product export is decreasing year after year. By 1980-81, the share dropped to 38.56 percent in quantity and 40.49 percent in value. By 2012-13 the share declined to 17.93 percent in quantity and 18.22 percent in value (MPEDA). This is mainly due to the growth of aquaculture in states like Gujarat, Andhra Pradesh, Orissa, West Bengal etc. However, the volume of export from Kerala is generally on an increasing trend.

Table 2.5: Export Trends of Marine Products- India & Kerala 2007-08 to 2012-13

Year	INDIA		KERALA		KERALA's Share (Percent)	
	Quantity (Tonnes)	Value(Rs Lakh)	Quantity (Tonnes)	Value (Rs Lakh)	Quantity	Value
2007-08	541701	762092	100318	143091	18.52	18.78
2008-09	602835	860794	100780	157218	16.72	18.26
2009-10	678436	1004853	107293	167002	15.81	16.62
2010-11	813091	1290147	124615	200210	15.33	15.52
2011-12	862021	1659723	155714	298833	18.06	18
2012-13	928215	1885626	166399	343585	17.93	18.22

Source:http://www.spb.kerala.gov.in/images/pdf/er13/Chapter2/chapter02.html# Fisheries Economic Review 2013, State Planning Board, Government of Kerala

Analysis of Export from India-Gravity Model

Gravity model examines the magnitude of trade between countries which cannot be explained by other theories of international trade. The classical and new trade theory can successfully explain the reasons for countries to join in world trade. However they cannot answer the question of the size of the trade flows. Another trade theory is the gravity model, which has been intensively used in analysing the performances of international trade in recent years and can be applied to quantify the trade flows empirically. The gravity model applies Newton's universal law of gravitation in physics which states that the gravitational attraction between two objects is proportional of their masses and inversely relates to square their distance. Timbergen was first who applied the gravity model to analyse international trade flows in 1962 and many others has followed to set up a series of econometric model bilateral trade flows. The general gravity model applied in the bilateral trade has the following form.

$T_{ij} = A\ Y_iY_j / D_{ij}$

Where A is a constant term

Tij is the total trade flow from origin country i to destination country j

Yi,Yj are the economic size of two countries i and j. Yi,Yj are usually gross domestic product(GDP) or Gross National Product(GNP).

Third, distance is another commonly used variable in gravity model. Distance is the geographical distance between countries. It is also a proxy for transport cost, which is usually measured as the straight line distance between countries economic centers. Table No.2.6 Presents the estimation of gravity model covering the period of 24 years from 1980 to 2014. Trade flows between two countries i and j are explained by factors that indicate total potential supply of country i, total potential demand of country j and the resistance factors to trade flow between i and j. The gravity model is then obtained by equality of supply and demand. Then the gravity equation is obtained by using market equilibrium clearance.

Table No.2.6: Gravity Analysis for Marine Exports from India and Importing Countries

Dependent Variable : Log of Marine Products Exports from I to j		
Independent Variable	**Coefficient**	**Standard Error**
Constant	-54.1485 (-6.43005)*	8.421165
Log GDPi	0.710515 (1.676091)***	0.423912
Log GDPj (Average)	3.95464 (3.732296)*	1.059573
Log Dij	-0.13551(-0.89161)	0.151977
R2 =0.950798Adjusted	*R2* = 0.945878	

Source : Computed from MPEDA Government of Kerala, Kochi., World Bank Database, 1980-2014

Note : * statistically significant at 1 percent

** Statistically significant at 5 percent level

*** Statistically significant at 10 percent level

Figures in parentheses Indicates t-Statistic Value

The Gravity model estimated in this study has the following form:

$T_{ij} = A\ y_iy_j / D_{ij}$

$\text{Log}(EV_{ij}) = C + a \log GDP_i + b \log GDP_j + c \log D_{ij} + U_i$

LogEVij = Log of Marine Exports from i to j (i =India j = import markets)

C = Constant

Log GDPi = log of India's Gross Domestic Product at constant prices

Log GDPj =log of importing countries Gross Domestic Product at Constant Prices (log of Average Gross Domestic Product at Constant Prices (USA,Japan,EU))

Log Dij = log of distance between exporting and importing nations
Ui =Error term

The result of the gravity model with respect to importing countries reveals that growth in GDP of India and importing countries has a positive influence on the marine fisheries export from India. Log GDPj of importing countries is significant at one percent level. However LogGDPi of India is significant only at 10 percent level. The value of R2 and adjusted R2 are found to be high.

Gravity Analysis for Marine Exports India and USA

Tij = A yiyj/Dij

Log Tij= log A + Log Yi+ Log Yj – Log Dij + Ui

Log T ij= Log of Marine Exports from *i to j* (I =India J = import markets*)*

Log A = Constant

LogYi = log of India's Gross Domestic Product at constant prices

Log Yj = log of USA's Gross Domestic Product at Constant Prices

Log Dij = log of distance between exporting and importing nations

Ui =Error Term

Table No.2.6a: Gravity Analysis for Marine Exports India and USA

Dependent Variable : Log of Marine Products Exports from i to j		
Independent Variable	**Coefficient**	**Standard Error**
Constant	-48.2852 (-6.87909)*	7.019122
Log GDPi (India)	0.641527 (1.420586)***	0.451593
Log GDPj (USA)	3.399963 (3.650011)*	0.931494
Log Dij	-0.22008 (-1.47564)***	0.149146
R2 =0.950107Adjusted *R2* = 0.945118		

Source : Computed from data collected from MPEDA Government of Kerala, Kochi., World Bank Database, 1980-2014

Note :* statistically significant at 1 percent

** Statistically significant at 5 percent level

*** Statistically significant at 10 percent level

Figures in parentheses Indicates t-Statistic Value

The result of the gravity model with respect to USA reveals that growth in GDP of India and USA has a positive influence on the marine fisheries export from India. Log GDPj of USA is significant at one percent level. However LogGDPi of India is significant only at 10 percent level. LogDij shows the right sign. However it is significant only at 10 percent level. The value of R^2 and adjusted R^2 are found to be high.

Gravity Analysis for Marine Exports India and European Union

Tij = A yiyj/Dij

Log Tij= log A + Log Yi+ Log Yj – Log Dij+ Ui

Log T ij= Log of Marine Exports from i to j (I =India J = import markets)

Log A = Constant

LogYi = log of India's Gross Domestic Product at constant prices

Log Yj = log of European Union's Gross Domestic Product at Constant Prices

Log Dij = log of distance between exporting and importing nations

Ui =Error Term

Table No. 2.6b: Gravity Analysis for Marine Exports India and European Union

Dependent Variable : Log of Marine Products Exports from I to j		
Independent Variable	**Coefficient**	**Standard Error**
Constant	-58.0008(-4.61182)*	12.57656
Log GDPi (India)	0.886748 (1.776979)***	0.49902
Log GDPj (EU)	3.904435 (2.796274)*	1.396299
Log Dij	-0.17616 (-1.08309)*	0.162647
R2 =0. 942847 Adjusted *R2* = 0. 937132		

Source : Computed from data collected from MPEDA.World Bank Database, 1980-2014

Note :* statistically significant at 1 percent

** Statistically significant at 5 percent level

*** Statistically significant at 10 percent level

Figures in parentheses Indicates t-Statistic Value

The result of the gravity model with respect to European Union reveals that growth in GDP of India and European Union has a positive influence on the marine fisheries export from India. Log GDPj of European Union is significant at one percent t level. However Log GDPi of India is significant only at 10 percent level. Log Dij showing the right sign. However it is significant only at 10 percent level. The value of R^2 and adjusted R^2 are found to be high.

Gravity Analysis for Marine Exports India and Japan

Tij = A yiyj/Dij

Log Tij= log A + Log Yi+ Log Yj – Log Dij+ Ui

Log T ij= Log of Marine Exports from i to j (I =India J = import markets)

Log A = Constant

LogYi = log of India's Gross Domestic Product at constant prices

Log Yj = log of Japan's Gross Domestic Product at Constant Prices

Log Dij = log of distance between exporting and importing nations

Ui =Error Term

Table No.2.6c: Gravity Analysis for Marine Exports India and Japan

Dependent Variable : Log of Marine Products Exports from I to j		
Independent Variable	**Coefficient**	**Standard Error**
Constant	-49.3682 (-8.42393)*	5.860471
Log GDPi	1.384581 (6.615095)*	1.384581
Log GDPj	2.905969 (4.572036)*	2.905969
Log Dij	0.048728 (0.31635)	0.048728
R2 =0. 957538 Adjusted *R2* = 0. 953292		

Source : Computed from data collected from MPEDA Govt. of Kerala., World Bank Database, 1980-2014

Note :* statistically significant at 1 percent

** Statistically significant at 5 percent level

*** Statistically significant at 10 percent level

Figures in parentheses Indicates t-Statistic Value

The result of the gravity model with respect to Japan reveals that growth in GDP of India and Japan has a positive influence on the marine fisheries export from India. Log GDPj of Japan is significant at one percent level. Similarly LogGDPi of India is significant 1 percent level. The value of R^2 and adjusted R^2 are found to be high.

Gravity Model has given in the Table No. 2.6a; Table No. 2.6b and Table No. 2.6c show that the growth in GDP has a positive impact on Marine fisheries export trend from India. In the model, the coefficient of the GDP remains at constant price is positive and highly significant. The distance variable is significant and has the anticipated negative sign, which indicates that India tends to trade, with its neighboring countries during the period of food safety regulations. This variable is expected to have negative effect on trade, as transport cost increase with the distance between countries. This negative coefficient shows the decline in Indian export and upward trend in the domestic consumption.

Four Phases of Fisheries Scenario

The fishing industry in Kerala is of vital importance in the economic development of the State. Due to the economic impact of rapid growth of population which leads to manifold effect of social and economic issues especially poverty and under-employment and diminishing economic returns, requisite for development of fisheries along with other industries as a means to resolve the problems has been greatly highlighted in recent years. This chapter traces the growth and contribution of fish production in the economy of India as well as Kerala and also analyses the exports of marine fish production in Kerala. Fisheries scenario has been categorized into four Phases which examines the profile, the production, export trends, destination changes, product diversification, technological innovations, recent challenges and issues explored.

Marine fisheries in India, including Kerala are currently passing through a crisis mainly due to stagnation in production, higher operational cost and low profitability. For an understanding of the real status of the fishing industry in India and Kerala, it is necessary to divide the period from 1960 to 2014 clustered into four phases. This division is based on the foremost issues faced by the fishing industry and can be explained on the basis of the compound annual growth rate of marine product exports and fish production from India and Kerala summarized in the Table No.2.7 Compound annual growth rate is calculated by using the following formula.

$Y = ab^{t}$

$\text{Log } Y = \text{Log } ab^{t}$

$\text{Log } Y = \text{Log } a + t \text{ Log } b$

$b = [\text{Antilog of Log } b - 1]100$

Compound Annual Growth Rate = $[\text{Antilog of Log } b - 1]100$

Y is the Fish Production, Quantity and Value of marine product exports from India and Kerala, t is the time period.

Table No. 2.7: Compound Annual Growth of Marine Export Trend and Fish Production from India and Kerala

PERIOD	CAGR FISH PRODUCTION		CAGR MARINE EXPORT			
	INDIA	KERALA	INDIA		KERALA	
			QTY Tonnes	Value Rs. (Crores)	QTY Tonnes	Value Rs. (Crores)
I PHASE 1960- 1985	10.9	2.1	9.2	24.6	7.1	18.8
II PHASE 1985-1997	6.3	7.2	16	26.9	6.3	20.5
III PHASE 1997-2004	2.8	1.2	4.7	6.3	-1.5	2.7
IV PHASE2004-2014	4.7	0.7	8.8	16.7	2.8	15.8
OVER ALL 1960-2014	5.2	2.1	8.4	17.3	5.2	12.9

Source: Computed from data collected from MPEDA Government of India, Kochi. Economic Review, Government of Kerala, Various Years. CMFRI, Government of India, Kochi.

The first phase covers the period 1960 to 1985 and is termed as slow modernization phase due to high demand for prawns from the international market. The second phase is from 1985 to 1997 and witnessed a rapid expansion period. This period faced the challenges of economic and ecological impact of motorization, exploitation of deep sea region, ban on monsoon trawling and the era of New Economic Policy with liberalization, privatization and globalization which stimulated the marine export trends that lead to changes in the product components and also the destination of the marine product exports. The third phase is from 1997 to 2004 and is observed as quality revolution period which is observed the European Union ban period, implementation of stringent international quality assurance standards and HACCP.

Quality concept has undergone tremendous changes in this period. Earlier the quality control was done at the end product stage. Quality control has changed into Quality Assurance which in the case of fish, starts right from the point of catch, ensuring hygienic handling and storage, onboard the fishing vessel. At every stage of handling, transportation, storage, preprocess handling, processing, packaging, storage of finished product to the end market, strictest standards of hygiene are to be ensured. Quality involves so many other aspects like the quality of water used for seafood freezing industry. EU countries insist on continuous monitoring which sternly demanded the establishment of good analytical laboratories, fully equipped with best trained and highly qualified technologist.

Thus human resource development has become a major need and has revolutionised the seafood processing export industry. Compared to the other phases, the compound annual growth rate of quantity and value of marine product exports from India and Kerala has drastically declined especially in the third phase. Kerala's marine export quantity turned even negative during this phase. The main reason for this has been the state's inability to keep up with Andhra Pradesh in shrimp culture. According to MPEDA statistics, of the Rs.4535 crores worth shrimp that was exported from India during the year 2000-2001, 86 percent came from shrimp culture. The major reason for the marginal increase in the value share has been the rising export of low frozen fishes such as ribbon fish and mackerels to China and Hong Kong.

Another reason for the decline is the problems in the 1990's, which was hit by disease and environmental concerns and the total output from this sector declined 18 percent during the decade. The fourth phase was from 2004 to 2014 which highlights the issues of antidumping duty on Indian shrimp and the effect of global recession on seafood export processing industry. Consequently, the study attempts to converse the various aspect of anti-dumping mechanism, and the various problems and issues faced by the seafood export processing industry. Above all the issues, the major concern was due to the raw material scarcity i.e. lack of availability of sufficient quantities of raw material at a reasonable price that resulted many processing industry to work at a low capacity utilization level.

The First Phase-Till 1985

In the present study, the first phase exposes early period till 1985. The growth of marine fisheries in India can be grouped into three periods .The first period extends from 1950 to 1960 witnessed a slow, but country has no steady growth in fisheries where non-mechanized craft and gears were operated. The second period from 1960 to 1980 was marked with use of improved gear materials, export trade expansion, increased mechanisation, initiation of motorisation of country craft and intensification of fishery activities. The third period which extends from mid 1980s to 1985 characterised by intensification of mechanised fishing, growth in motorization and multiday fishing, extension of fishing grounds, seasonal fishery ban, introduction of molluscan aquaculture, open sea cage farming and breeding, acoustic fish detection and satellite-based remote sensing techniques, advances in electronic navigation, provisions for on-board fish processing and preservation.

Fish Production and Export Trend

In pre-independence India, fisheries sector did not receive much attention from the government. Poor fishermen in their canoes and country crafts fished in the coastal and near shore waters using their cotton twin nets, catching sardine, mackerel, silver bellies, anchovies, Bombay duck, etc. Small quantities of prawns landed were dried into crude prawn pulp and exported to nearby countries like Burma. The sardine and mackerels were Colombo-cured and sent to the estate Labour in Northern Sri Lanka. An MPEDA Study observed that after the World War II Sri Lanka, Singapore and Malaysia slashed their imports of Marine Products and this posed a disastrous threat to India's Marine Products export.

Table 2.8: Growth in Export of Marine Products I Phase

Year	Qty in Tonnes	Value in Rs. Crore	Average Unit Value Realiza-tion (Rs. / Kg)	Value in US $ Million	Average Unit Value Realiza-tion US $ / Kg.	Growth Rate % Qty	Growth Rate %Rupee Value	Growth rate %Dollar Value
1961-62	15732	3.92	2.49	NA	NA	-21.3	-15.52	NA
1962-63	11161	4.2	3.76	NA	NA	-29.06	7.14	NA
1963-64	19057	6.09	3.2	NA	NA	70.75	45	NA
1964-65	21122	7.14	3.38	NA	NA	10.84	17.24	NA
1965-66	15295	7.06	4.62	NA	NA	-27.59	-1.12	NA
1966-67	21116	17.37	8.23	NA	NA	38.06	146.03	NA
1967-68	21907	19.72	9	NA	NA	3.75	13.53	NA
1968-69	26811	24.7	9.21	NA	NA	22.39	25.25	NA
1969-70	31695	33.46	10.56	NA	NA	18.22	35.47	NA
1970-71	35883	35.07	9.77	46.4	1.29	13.21	4.81	NA
1971-72	35523	44.55	12.54	59.61	1.68	-1	27.03	28.47
1972-73	38903	59.72	15.35	77.81	2	9.51	34.05	30.53
1973-74	52279	89.51	17.12	114.87	2.2	34.38	49.88	47.62
1974-75	45099	68.41	15.17	86.15	1.91	-13.73	-23.57	-25
1975-76	54463	124.53	22.87	143.43	2.63	20.76	82.03	66.48
1976-77	66750	189.12	28.33	210.66	3.16	22.56	51.87	46.88
1977-78	56967	180.12	31.62	209.79	3.68	-14.66	-4.76	-0.41
1978-79	86894	234.62	27	285.19	3.28	52.53	30.26	35.94
1979-80	86401	248.82	28.8	307.28	3.56	-0.57	6.05	7.74
1980-81	75591	234.84	31.07	296.92	3.93	-12.51	-5.62	-3.37
1981-82	70105	286.01	40.8	318.91	4.55	-7.26	21.79	7.41

Year	Qty in Tonnes	Value in Rs. Crore	Average Unit Value Realiza-tion (Rs. / Kg)	Value in US $ Million	Average Unit Value Realiza-tion US $ / Kg.	Growth Rate % Qty	Growth Rate %Rupee Value	Growth rate %Dollar Value
1982-83	78175	361.36	46.22	373.85	4.78	11.51	26.35	17.23
1983-84	92187	373.02	40.46	360.75	3.91	17.92	3.23	-3.5
1984-85	86187	384.29	44.59	323.24	3.75	-6.51	3.02	-10.4

Source MPEDA, Government of India, Kochi.

The period also witnessed worldwide demand for a new item frozen prawn especially from Japan and Western Countries including USA. The modernisation of Indian seafood industry began after the country's independence in 1947. Besides, the post-independence days the importance of fisheries development was realized and coastal and maritime states, especially along the south west coast, registered a phenomenal increase in annual fish landings.

Fisheries sector has witnessed a steady growth from the first five year Plan onwards with the annual fish production increasing from around 0.53 million tonnes in1950-1951 to 9.65 million tonnes during 2013-2014. Table 2.8 and the Figure 2.2 depict the growth in Export of Indian Marine Products from 1961 to 1985. Starting from a purely traditional activity in the fifties, both aquaculture and fisheries have transformed to commercial enterprises opening considerable potential for employment generation and contribution to the food and nutrition security and foreign exchange earnings of the country(Planning Commmission,2001). Japan, the EU and the US import a major proportion of India's exports, and the main export species is shrimp.

Kerala is not only the highest producer of marine fish but also biggest consumer too. Fisheries sector is an important source of foreign exchange. Increases in fish production during the first four decades of planned development in the country came from the marine sector, especially the coastal waters (Planning Commission, 2001). Kerala occupies the foremost position in marine fish production in India, accounting from 1.96 lakh tonnes (average of 1950 and 1951) to 3.90 lakh tonnes (average of 1983 and 1984) about twice increase in production while, at the all-India level, production has tripled. Prior to 1956, the catch was of the order of 6000 tonnes which rose to 14,000 tonnes by 1959. By mid-sixties it was of the order of 25,000 tonnes which doubled (51,000 tonnes) during 1970-74. The average fish production during 1958-59 to 1960-61 was about 3.2 lakh metric tonnes of which 2.7 lakh tonnes were from the sea while the rest came from the inland waters. The total quantity of all marine products exported from Kerala increased from 24,000 tonnes in 1970 to 33,000 tonnes in 1982 in the corresponding period the value of exports increased from Rs. 275 million to Rs. 1380 million (Sebastian Mathew,1986). Between 1957-58 and 1983-84 the share value of the marine exports to Kerala's total export value increased from 2.6 percent to 21.5 percent. During these periods fish is treated as the cheapest source of animal protein lavishly available in Kerala. Figure 2.3 shows the marine fish landings in India and Kerala for the period 1962 to 1985.

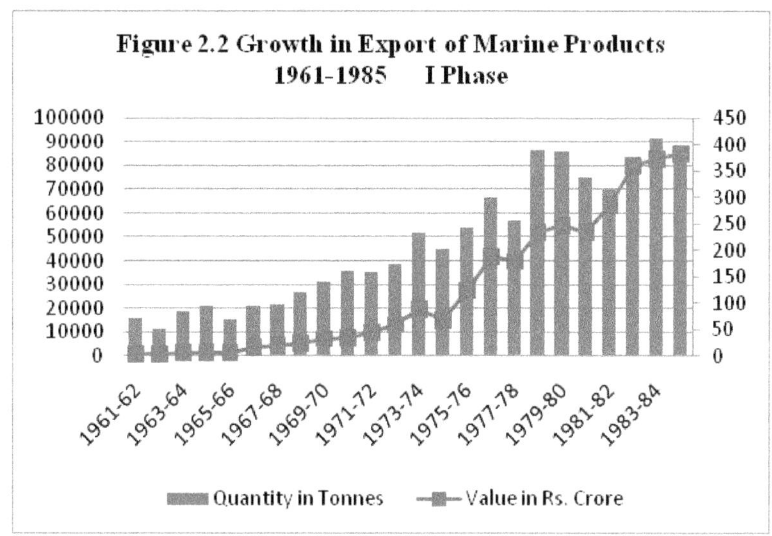

Source: MPEDA, Government of India, Kochi.

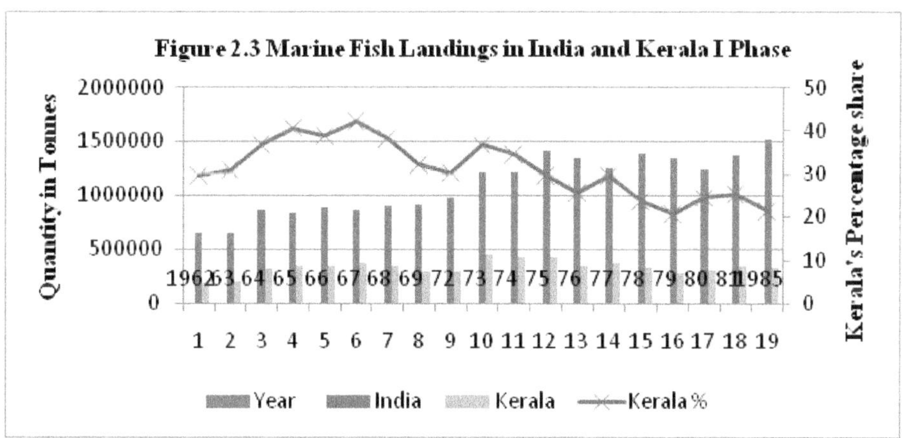

Source: CMFRI, Government of India, Kochi.

During this phase Kerala's marine fish landing share decreases from 30 percent to 21 percent. Declaration of Indian EEZ was the major reason for this. Annual pelagic fish landings are estimated at 2.64 lakh tonnes (average of 1983 and 1984) and the demersal 1.26 lakh tonnes, the former accounting for 68 percent and the latter 32percent. Maximum fish landings, of about 93,000 tonnes (average of 1980 to 1984) were observed in Quilon district, accounting for 28 percent of the total landings of the state. Trivandrum comes next with a catch of 48,000 tonnes (accounting for 14 percent, followed by Ernakulam 38,900 tonnes (12 percent) Alleppey 37,800tonnes (11 percent), Cannanore 36,400 tonnes (11 percent) Kozhikode 28,400 tonnes (9 percent), Malappuram 26,100 tonnes (8 percent) and Trichur 23,300 tonnes (7 percent).The prawn catches have been steadily increasing since mid-fifties till mid-seventies.

Additional production of prawns and their exports have significantly contributed to the national income and thus foreign exchange earnings had a phenomenal growth. Thus even though total production did not show any marked improvement, due to the positive impact of the shrimp fishery, the fishery sector economy prospered.

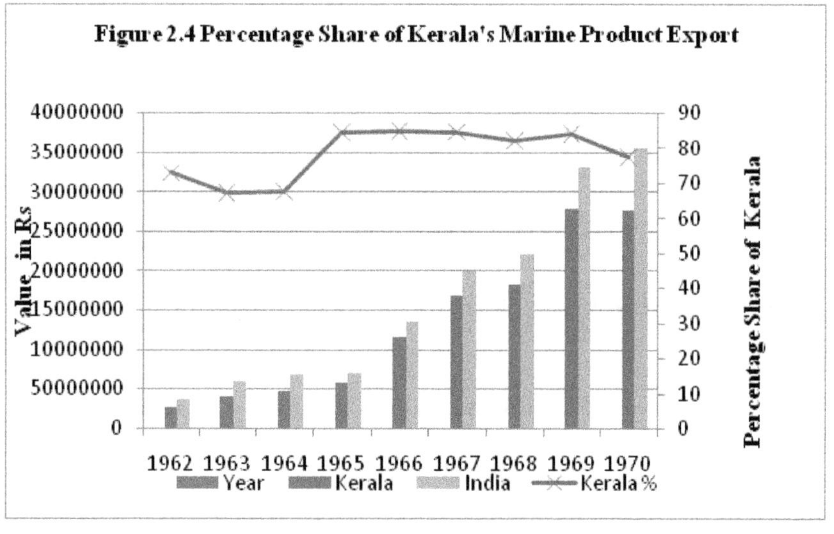

Figure 2.4 Percentage Share of Kerala's Marine Product Export

Source: MPEDA, Government of India, Kochi.

The value of the marine fish landings (1983) of Kerala at the landing centre price has been estimated at about Rs. 114 crore. About half the total value realised is accounted by prawns. The annual per-capita availability of fish in fresh form has been estimated at about 13 kg. This is about 3 times the availability at the all India level. However there is need to step up production to meet the demand of the comparatively large proportion of fish eating population of the state. Figure 2.4 shows Marine Product Export trend in India and Kerala during 1962 to 1970. During this period Kerala was the chief supplier of seafood export.

Technological Innovations

The economies of fishing operation depends on the market condition which varies with the technology adopted and size, range of fishing crafts(A V Shibu,1999). Organised fishing using mechanical propulsion in India started only in the early fifties although sporadic attempts of explorating fish surveys have been made by the turn of this century (Varghese C P, 1994). The major projects introduced in

Kerala, aided by foreign agencies, were Technical Co-operation Mission Programme (TCMP), FAO Technical Assistance Programme (FAO-TAP) and Indo-Norwegian Project (Rajan B J, 2002). Mechanisation was started in late fifties by the Indo-Norwegian Project, at first in the Quilon district of Kerala. The Fishermen Training Centre established with the assistance of FAO at Satpati in 1954 was a forerunner of a chain of such centers in all the maritime states of India to train fishermen in operating mechanized fishing boats (Silas E G, 1977). The fishing industry provides employment to over 2 lakhs of which fishermen alone number about 83,200. The fishing equipment consists of nearly 12,000 big and small non-mechanised canoes, over 8,000 catamarans and thousands of hooks and lines. The new technology of craft and gear has resulted in overcapitalization.

Experiments done by the Indo-Norwegian Project and the F.A.O. have proved that the existing indigenous crafts are not suitable for mechanization. Their range of operation can be developed under what is called dory fishing and mother-ship operation by powered vessels towing them to and from wider fields of operation. The Indo-Norwegian Project boats have been issued to fishermen on subsidized costs. The subsidies have been 25 percent on hull, 50 percent on engine and 331/2 percent on gear. The cost is recovered in 40 easy installments spread over a period of 5 years. The boats issued by the state department have been on hire purchase system to groups of 4 to 6 fishermen trained from the Fishermen Training Centers at Beypore and Ernakulam. The subsidized cost has to be covered in easy terms of 30 percent of the daily catches.

During the early sixties there is a shift from cotton to nylon and it paved the way for technological progress and by mid-sixties a sudden advancement of trawl fishery in the coastal waters. The new fishing nets introduced was Nylon drift and gill nets and small trawls. During 1960-61, 2,011 metric tonnes of fish valued at Rs. 6.69,643 were landed from the mechanized boats. Of this, 1086 metric tonnes have been landed at Neendakara, 453 at Cochin, 252 at Azhicode, 70 at Tellicherry, 135 at Beypore, 10 at Ponnani and 5 at Vizhinjom. Commercial purse-seine started during the late seventies and the process of large scale motorization of country crafts began in early eighties (Jacob T et al., 1987). In the beginning of 1973 the Government established the Marine Product Development Authority replacing the Marine Export Promotion Council established in 1961. The total real income from fishing accruing to the traditional fishermen community as a whole has increased by hardly 20 percent between 1969-71 and 1974-76 and in per capital terms it was only 7 percent (John, 1978). At current prices the total share of the fishermen working in the mechanized fishing sector has risen by about 500 percent compared to 77 percent in the non-mechanised sector during the same period indicating a widening disparity in incomes between the fishermen of the mechanized and non- mechanized sectors.

Deep sea fishing and development of off-shore fishing are outside the scope of the small mechanized crafts and are pursued by the Indo-Norwegian Project. Lady Goschen in 1927-30 and Chandrika and Sagarkumari in 1949 have conducted investigations led to the discovery of the rich shark fisheries of the south that finally paved the way for the development of the shark liver oil factory. In 1981 an organized section of the artisanal fishermen spearheaded a movement demanding measures to regulate the anarchic and destructive fishing by trawlers. They demanded a monsoon trawl ban. This was highlighted as essential to prevent further marine resource depletion in the inshore waters. The commencement of commercial purse-seine fishing in 1979 in Kerala was opposed by the artisanal fishermen and it led to clashes between the artisanal and the pursue operators which created tensions on many occasions (Nair et al., 1983). Outboard driftnet and gillnet operations started in Kerala during 1984. During 1985-97, a heavy landing was recorded and ring seine operations picked up momentum. There was a significant increase observed in the Kerala's marine fish landings since the up-gradation and mechanization of country crafts in the state and the landings from the non-mechanised sector become very nominal. Oil sardine, Indian mackerel, penaeid prawns, carangids, perches, cephalopods, tunnies and ribbon fishes were some of the major commercially important resources.

Destination Changes and Product Diversification

Between 1950 and 1960 India's exports was dominated by traditional items like fish oil, salted and dried fish, dried shrimp, shark fins and fish maws India exported to traditional markets such as Sri Lanka, Burma, Singapore, Malaysia and Hong Kong. This situation changed with the development of technology and modernization, dried products gave way to canned and frozen items. The product shift also resulted in market shift. Initially, canned shrimp exports were mainly focused and then due to non-availability of suitable cans in the country, the industry was shortly compelled to move to exports of frozen shrimp. Since 1961 the export of dried marine products was overtaken by export of frozen items leading to a steady progress in export earnings. With the devaluation of Indian currency in 1966 the export of frozen and canned items registered an increase and led to more processing plants mushrooming across the country. The export of other varieties of Fish like Squid, Cuttlefish, Octopus, Crabs, Clams and Mussels started later.

Frozen items continued to dominate the trade. Markets for Indian products also spread fast to developed countries from the traditional buyers in neighbouring countries. More sophisticated and affluent markets viz. U.S.A, Japan, U.K. Australia, France, Ceylon etc. became our important buyers for a long. This is depicted in Table 2.9. USA was the principal buyer for our frozen shrimp but after 1977, Japan emerged as the principal buyer of the product, followed by the West European countries. Several seafood processing units with modern machinery for freezing and production of value-added products were set up at all important centers in the country for export processing. In order to meet the growing demand of capture fisheries and to hinder exploitation MPEDA pioneered the development of coastal aquaculture.

Table 2.9: Major Importers of Indian Marine Products and Their Share in Our Exports. Q: Quantity, V: Value

Major Importers		1963	1964	1965	1966	1967	1968	1969	1970
U.S.A	Q:	28.3	31.1	43.4	42.4	42.9	48.4	55	47.8
	V:	46.3	48.8	57.7	59.9	48.9	54.3	54.8	47.3
Japan	Q:	0.5	3.6	5.8	6.2	12.8	13.7	17.5	18
	V:	0.9	7.3	6.8	9.3	21.3	22.5	27.9	29.9
U.K.	Q:	1	0.8	2.2	2.8	3.9	3.4	2	2.5
	V:	2.9	2.1	3.9	4.4	5.5	4.2	2.5	4
Australia	Q:	1.5	2.9	3.6	3	2.8	2	2.6	2
	V:	3	5.7	5.5	4.5	4.2	2.7	3.6	3.1
France	Q:	0.6	0.4	2.5	3.2	3.4	2.1	1.8	1.5
	V:	1.6	1	4.9	5.6	5.3	2.8	2.4	2
Ceylon	Q:	47.8	49.8	29.9	34.8	23.8	21.2	13.5	19.2
	V:	22.4	24.4	9.7	10	6.6	6.1	3.3	5
Total	Q:	79.7	88.6	87.4	92.4	89.6	90.8	92.4	91
	V:	77.1	89.3	88.5	93.7	91.8	92.6	94.5	91.3
Other Countries	Q:	20.3	11.4	12.6	7.6	10.4	9.2	7.6	9
	V:	22.9	10.7	11.5	6.3	8.2	7.4	5.5	8.7
Total	Q:	100	100	100	100	100	100	100	100

Source: Statistics of Marine Products Exports 1970, Marine Product Export Promotion Council, Kochi.

The total output of fish in the country in 1965 was 6, 15,120 metric tonnes, an increase of 1, 00,000 tonnes over the previous year. It is estimated that 50 percent of the annual production of fish consumed is fresh while the rest is dried or preserved, canned and converted into products such as sauce for local consumption with rice as well as for export.

In the seventies, the export was depending mainly on shrimp but due to the export promotional measures, it became possible to diversify the products in the eighties adding cephalopods(cuttlefish, squid,, and octopus) and frozen fish (such as pomfrets, ribbon fish, seer fish, mackerel, reef cod, croakers, snapper etc). While all these items hold good prospects, live fish, chilled fresh water fish etc. are promising items for the future (Ayyappan and Krishnan 2005). India's first seafood Export to Japan was in September 1972 and towards the end of 1970s, Japan was the home of the India's biggest market share. There were changes in the demand for differentiated products from various countries. Japan indicated their preference for headless shell on shrimp; the USA demanded peeled shrimp meat and the European Countries preferred the IQF shrimp in frozen or cooked form and a major share of cephalopods. Frozen fish items had greater demand in the South East Asian countries as well as the Middle East (SaradaC et al., 2006). The principal marine fisheries of Kerala are sardine, mackerel, sole, prawn, cat fish, silver belly, white bait, shark and ray, ribbon fish, big-jawed jumper, tunny and seer fish. Figure 2.5 shows Indian marine products in 1984. During the eighties the canned items slowly disappeared and frozen items gained prominence in India's seafood trade. Amongst the frozen items also, there was changes in the demand for differentiated products from various countries. While Japan showed their preference to headless shell on shrimp, the USA demanded peeled shrimp meat and European countries preferred the IQF shrimp in frozen and cooked form.

Figure 2.5 Item wise Marine Products from India 1984(Quantity)

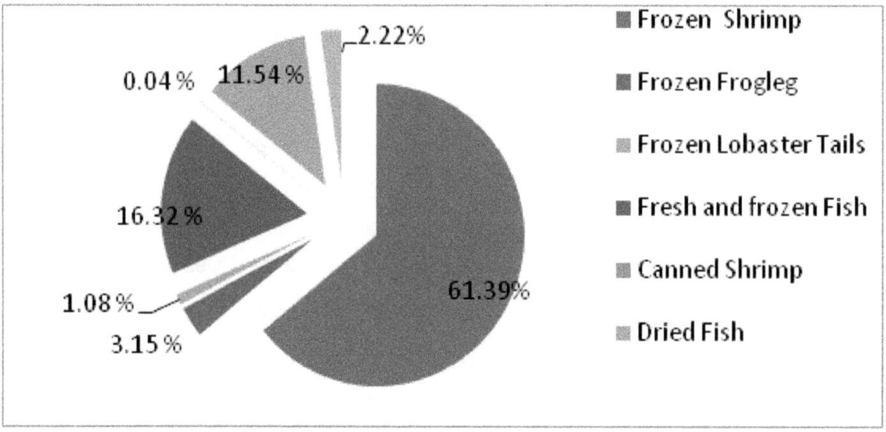

Source: Statistics of Marine Products Exports 1988, Marine Product Export Promotion Council, Kochi.

Infrastructural Facilities

At the beginning of the 1950s there existed only a very few ice factories in Kerala State, one at Calicut and three at Tellicherry, all privately owned factories. The Government had established a few ice factories which provide facilities for the production of 60 tonnes of ice per day with cold storage capacity for 235 tonnes of fish freezing capacity for 22 to 24 tonnes per day and freezing storage capacity for 775 tonnes of fish. The facilities in the private sector are over 40 tonnes per day for the production of ice with cold storage capacity for nearly 140 tonnes of fresh fish, freezing capacity for 14 tonnes per day and frozen storage capacity for 475 tonnes of fish. The improved facilities for Ice Making ,quick freezing plants with cold storage during 1950-60 have not only bettered the quality of the fish made available in the interior markets but also promoted a brisk industry in export trade of frozen and canned prawns, lobster tails and frozen frog legs with a number of foreign countries. There are 10 canning factories in the State. Nearly 10 percent of the marine fish catches is converted as fish meal, fish manure, etc. Many of the handling centers for manure, oil, and meal and prawn powder are very small units managed by individuals scattered all over the coastal belt. In Malabar section alone, there are 60 to 70 fish oil and guano factories. These by-products enjoy very good markets both internal and external. Items like prawn shell or powder have attracted a fairly good foreign market. In 1959-60, about 1,374 tonnes of prawn shell was exported to U K, Holland and other foreign countries.

The Government of Kerala has set up State Export Trade Development Corporation in 1984. The State Legislature has passed the Kerala Fishermen's Welfare Fund Act on 19th September 1985. The Kerala State Co-operative Federation for Fisheries Development Ltd. (Matsyafed) was established on the 19th of March 1984. Registered as a Co-operative Apex Federation of primary level cooperative societies, this organization has never looked back and is still growing incessantly. Fishing industry is subjected to various kinds of externalities and is influenced by technological as well as socio-economic development of Kerala economy. Mechanisation of fishing craft has brought about favourable change in the standard of living of the fishermen. Labour productivity in the mechanised sector is nine fold to that of the non-mechanized. A 36-fold increase in the value of the output is noticed between 1964 and 1984(Rajasenan, 1987). But this was mainly due to the very high unit price realisation of the export species of prawn.

The Research and Development effort for fishing harbours in the State was started in the sixties. The effort was to develop a number of minor and major fishing harbours in the state with central assistance. The programme was delayed till the beginning of the fourth five year plan in 1969 due to technical and other constraints. Even by the middle of the eighties the state had only one major fishing harbor at Cochin and five minor fishing harbors at Vishinjam, Azhicode, Ponnani, Baypore and Baliapattam. The work for a number of landing centers started during the sixth five year plan. Motorisation resulted in the introduction of outboard crafts led to new technological leap in traditional fishery in the mid 1980s. Due to the adverse effects of modernization on the traditional sector, the traditional fishermen realized the need for a new technology in the traditional sector as a survival strategy and this resulted in the introduction of outboard motors which become the major production unit in Kerala fishery with its average contribution being 38.4 percent more than the mechnaised or other traditional sector. During this period a quantum jumps in the output is due to modernization of traditional fishing units.

The Second Phase: Rapid Expansion Period from 1985 to 1997

The year 1985-86 has seen to boost economic growth, reinforce anti-poverty programmes, revitalize industry and provide a rapid expansion path to fisheries trade. The Indian Seafood export has shown an increasing trend during the II Phase. Socio-cultural, techno-economic and ecological parameters that affected a specific case of technology diffusion in marine fisheries (John, 1994). Diffusion has been defined as the process by which an innovation is communicated through certain channels overtime among the members of a social system (Rogers, 1983). During this phase Kerala's fishermen played a development role and successful dissemination of the new technology. The successful diffusion of an innovation in a traditional community is a strong indicator that the community has socio economically and culturally understood and approved the innovation according to its own norms. The empirical analysis showed that India's exports have turned around in the mid 1980s after growing at low or moderate rates till then and the markets for India's exports diversified with Asian developing countries emerging as major destination after 1985-86 (Sarkat, 2004).

Fish Production and Export Trend

Fish production levels have increased from 2.8 million tonnes of fish and shell-fish in 1985-86 to 5.3 million tonnes in 1996-1997. During 1985-86 and 1996-1997, the marine and inland fish production levels have increased with an average annual growth rate of 2.68 percent and 8.06 percent per annum respectively. The share of inland fishery sector, which was 29 percent in 1950- 51, has gone up to about 50 percent in 1999-2000. Figure 2.6 shows the marine fish landing in India and Kerala for the period of 1985 to 1997.Kerala's share percentage in total marine fish landings declined from 27.3 percent in 1977 to 18.39 percent in 1987 and then it increased to 25.59 percent by 1993.

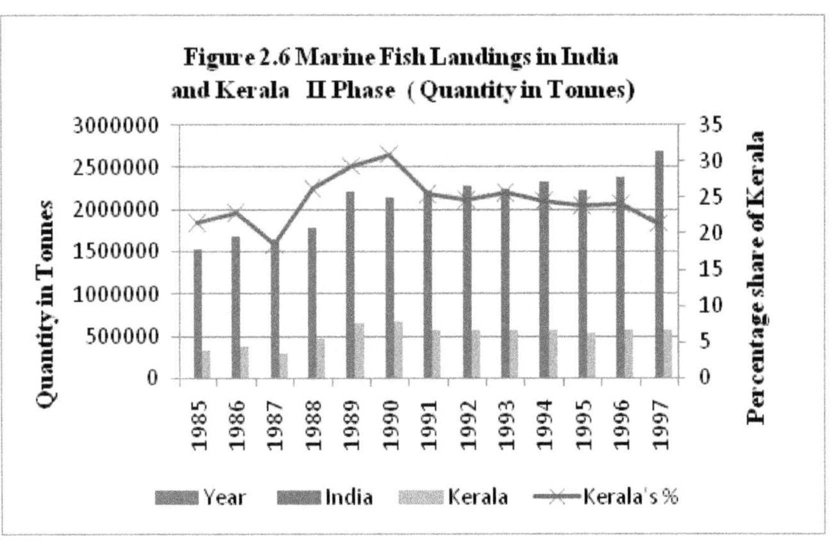

Source: CMFRI, Government of Kerala, Thiruvananthapuram & MPEDA, Government of India, Kochi

From Eighth plan onwards, the government has been encouraging fish seed production in the private sector. In the process, fish seed production has increased from 6322 million fry in 1985-86 to 15852 million fry in 1996-1997 (Appendix 2). Agricultural exports used to be of the order of 30.6 percent of the total exports during 1980-81, which came down to 19.4percent by 1990-91. More than 20percent of the marine products exported from India are from Kerala and the forex earnings by the state during 1994-95 were Rs. 816.64 crores. The need for giving emphasis for product development and value added products has been stressed to improve our forex earnings. Almost all type of fishing units operating along Kerala coast is found to be economically viable. Growth Rate of fish production has been a robust 4.12percent per annum during the nineties.

Kerala's fish production and export trend weaken steadily due to compliance with WTO regime, stringent regulations from the US Food and Drug Administration, EU regulations, social and economic issues. During this phase pre-shipment inspection was applied and teams of inspectors from the US visited Kerala to ensure sanitary conditions in prawn and fish processing factories.

The Economy of Kerala has to confront anti-competition and anti-liberalisation and opposition for opening up the world economy and global integration. More than 20 percent of the marine products exported from India are from Kerala. The growth in marine fisheries during the 1980s and 1990s has been slow as compared to the previous two decades (Planning Commission, 2001). During the last decade of 1990s the marine fish production has reached a plateau in Kerala. All over the world, increase in aquaculture production has been possible through diversification of aquaculture at different levels. However, diversification of Indian aquaculture is still at initial stages, inspite of the fact that India, producing 2.03 million tonnes (in 1998 according to FAO) is second to China (producing 69 percent of the world total). The disparity to a large extent is owing to the high level of diversification of Chinese aquaculture, which utilises several diversified aquaculture systems, seven major bio-categories (commodities) and a total of 29 species. Comparatively, India uses only two bio-categories (freshwater fishes accounting for 97 percent of production, and shrimps) and about 15 species.

The year 1995 turned out to be an important milestone in the history of India's seafood industry with shrimp production through aquaculture reached 100,000 Metric Tonnes. In the same year yet another milestone was crossed when marine product export crossed one billion US Dollars. But along with these laurels, the 90s brought new challenges. Shrimp farms in Andhra Pradesh, the largest shrimp producing State of India were affected by viral disease. Coastal aquaculture farming was sounded its death knell in 1996 when the supreme court of India following public interest litigation ordered the demolishing of all aquaculture farms set up within the coastal regulation zone. MPEDA's timely intervention through obtaining stay, subsequently the aquaculture authority of India was constituted to grant licenses to aquaculture farms within the coastal area. Growth rate of fish production has been about 4.12 percent per annum during the 90s. However, the full potential of all the water bodies have not been realized. They have not yet been able to utilise the full potential of inland fishery resources as well as the deep-sea resources. Inspite of sizable investment for marine fisheries in the past the growth of marine fish production in 1990s has been rather slow at an average of 2.19 percent as compared to the inland fish production growth rate of 6.55 percent.

Trend in export of marine products from 1985-86 to 1996-1997 is shown in Table 2.10. The quantity of marine products from the level of 83651 tonnes in 1985-86 has increased to 378199 tonnes in 1996-97, while the value of the export quantity has increased from Rs. 398 crores to Rs. 4121.36 crores during this period. In 1996-1997, the export has increased with an annual growth rate of 27.65 percent and

17.72 percent in quantity and value respectively. The volume and value of exports has increased over the years but average unit value realization (US $ / kg) has not shown any significant improvement. It showed a stagnant trend in this phase. This may be due to lack of value addition in the export composition. A study was carried out in Kerala coast during 1995-96, to assess the sectoral contribution of marine fish production, export trend of marine products and economics of prominent craft gear combinations. Production trend clearly indicates a declining catch of oil sardine and catfish in recent years and marked improvement in the production of penaeid prawns and cephalopods. More than 20percent of the marine products exported from India are from Kerala and the forex earnings by the state during 1994-95 were Rs. 816.64 crores. The need for giving emphasis for product development and value added products has been stressed to improve our forex earnings. Almost all type of fishing units operating along Kerala coast are found to be economically viable (Sathiadhas and Raghu, Kanakkand and Harshan, 2000). Kerala is a leading maritime state. Marine fish production increased from 325000 tonnes from the mid 1980s to 600,000 tonnes in the early 1990s.

Table 2.10 Growth of Marine Product Export from India 1985-1997 II Phase

Year	Qty in Tonnes	Value in Rs. Crore	Average Unit Value Realiza-tion (Rs. / Kg))	Value in US $ Million	Average Unit Value Realiza-tion US $ / Kg.	Growth Rate Percent Qty	Growth Rate percent Rupee Value
1985-86	83651	398	47.58	325.3	3.89	-2.94	3.57
1986-87	85843	460.67	53.66	360.51	4.2	2.62	15.75
1987-88	97179	531.2	54.66	409.69	4.22	13.21	15.31
1988-89	99777	597.85	59.92	412.83	4.14	2.67	12.55
1989-90	110843	634.99	57.29	381.39	3.44	11.09	6.21
1990-91	139419	893.37	64.08	497.9	3.57	25.78	40.69
1991-92	171820	1375.9	80.08	562.19	3.27	23.24	54.01
1992-93	209025	1768.6	84.61	610.63	2.92	21.65	28.54
1993-94	243960	2503.6	102.62	798.21	3.27	16.71	41.56
1994-95	307337	3575.3	116.33	1138.62	3.7	25.98	42.8
1995-96	296277	3501.1	118.17	1111.46	3.75	-3.6	-2.07
1996-97	378199	4121.4	108.97	1152.83	3.05	27.65	17.72

Source: Source: MPEDA, Government of India, Kochi.

Destination Changes and Product Diversification

Fish is the most important source of protein for many around the globe. Estimates are that, globally, per capita consumption of fish is 14.3 kilograms (kg) per year (Delgado et al., 2003). Per capita consumption in 1997 was led by Japan, with 62.6 kg per year and China at 26.5 kg per year, up from 8.1 in 1985. The European

Union (EU) consumes at 23.6 kg per year per capita and Southeast Asia at 23.0 kg per year, up from 19.8 in 1985. Furthermore, per capita consumption of fish by 2020 is expected to rise to 35.9 kg/year in China and 25.8 kg/year in Southeast Asia, while it will remain constant or decline in developed countries (Delgado et al., 2003). The direction of trade has also undergone changes since 1985 .With increased world demand for frozen prawn the export of marine product from Kerala showed greater market concentration. The marine product export industry has been capable of adapting to the need of the constantly changing demand pattern in the international market. Our export to Japan has increased from US$251.49 million in 1987-88 to US $ 641.68 million in 1997-98. In the comparable period export to Europe has increased from US$60.76 million to $113.80 million.

Steady growth has been seen due to positive rapid expansion which is taking place in the production front through shrimp farming and introduction of several resource specific vessels to enlarge the marine fish landings. Figure 2.7 shows destination of seafood in 1997.The increase in the price and foreign demand for prawn prompted the entry of large number of capitalist merchants into the industry. Dried fish was replaced by frozen prawn which was again replaced by IQF prawns and other value added shrimp products later.

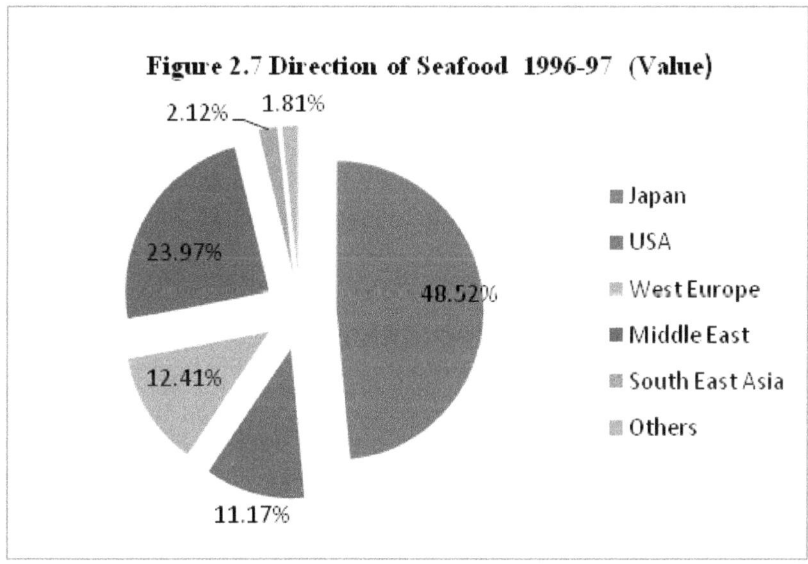

Figure 2.7 Direction of Seafood 1996-97 (Value)

Source: MPEDA, Government of India, Kochi.

Fishery resources can be utilised for the production of value added marine products for export and domestic market front now. Cuttlefish, squid products, live items are emerging as important item of exports. During this Phase marine products were exported to other countries also. Western Europe and South East Asian countries emerged as new importers of marine products from Kerala. The destination in percentage volume between 1991- 1997 depicts in the Figure 2.8 expose initial trend divergence in the GPL global market.

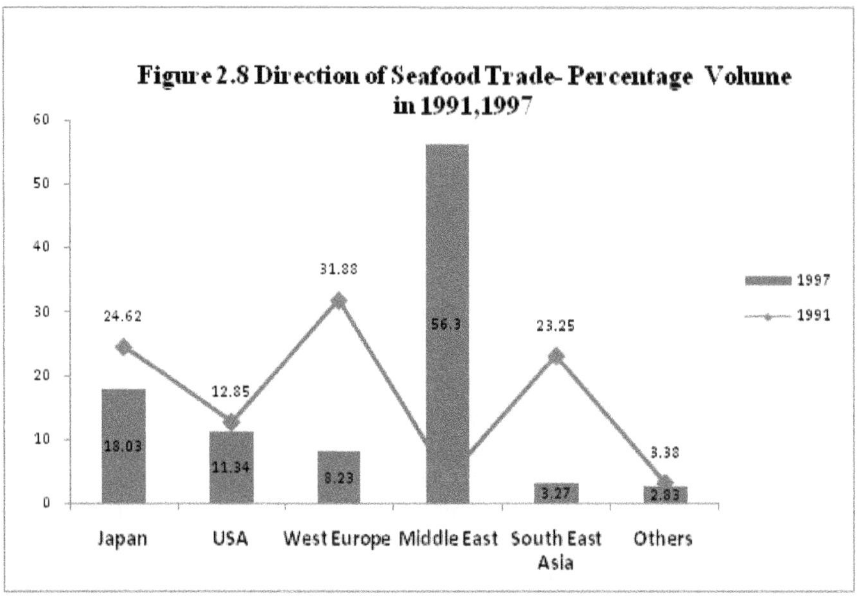

Source: MPEDA, Government of India, Kochi.

During the mid 1980's trade relations enhanced and also helped to find new markets for Indian marine products. Trade fairs, Indian seafood workshops, delegators foreign visits to Latin American Countries, South East Asia, Japan, West Europe, USA and Australia and inviting experts to India strengthening existing trade relations and sustaining the level of exports. Regular visit from buyer from foreign market also helped promoting Indian seafood trade.

Technological & Quality Revolution

The period of 1985 witnessed further progress in the marine fisheries development in the country. This period was characterized by rapid growth in motorising the artisanal fleet, further extension of fishing offshore and extended voyage fishing, and introduction of seasonal closures of selected fisheries as concerns developed over depleting fish stocks. Technology induced intensification of fishing effort and quantum leap in fishing efficiency, introduction of multi-day fishing and motorized craft and ring seins was the hallmark of this period. Maritime states enacted Marine Fishing Regulation Act (MFRAs). During the 1990s economic liberalization gave the fishing industry a major investment boost. There were 225 deep sea vessels operating in the Indian EEZ. Modernisation and technological advancement of seafood industry paved the way for higher exports. Government provides subsidy to boat owners for installing fish finders, GPS, Radio, telephone gadgets, etc. During this decade setting up a seafood park with world class preprocessing and laboratory facilities came true. The first batches of Indian seafood technologists were sent to USA to join hands with joint venture holding company named Marine Product Infrastructure Development Corporation Private Limited,

MIDCON. This joint venture company was registered as the Seafood Park India Limited (SPIC). SPIC is the first of its kind in Public Private Partnership.

Since 1990s, three issues dominated Indian export scene: decline in overall catches, particularly shrimp; fluctuations in international markets depressing prices and profitability; and overcapitalization of the production and marketing activities and increasing risk. In marine sector, inshore waters have been almost exploited to the Maximum Sustainable Yield (MSY) levels; the contribution from the deep sea has been insignificant. Inspite of sizeable investment for marine fisheries in the past, the growth in marine fish production in 1990's has been rather slow at an average of 2.19 percent as compared to the inland fish production growth rate of 6.55 percent. Indian Seafood export processing Industry was facing threats when a few consignments were rejected by the EU for detecting the presence of Salmonella. The final blow came after a European veterinary team visited processing plants in India and declared them to be non-compliant with EUs quality standards. Hence in 1996 a ban was imposed on export of marine products from India to Europe.

Infrastructural Facilities

Capital investment in the sector has more than doubled after 1988-89. In the mechanized sector, additional investment was used for increasing the number of boats, whereas in the artisanal sector it was mainly used for the purchase of higher horsepower engines and larger crafts. Though the marine fish production has considerably increased over the years, the economically backward traditional fishermen in the non-mechanized sector have been highly marginalized with enhanced mechanisation and motorisation of the fishing units. Inspite of the increase in fleet size and decrease in the catch rates, the mechanised and the motorised sectors still sustain mainly due to the increase in the price of almost all the varieties of fishes. Most of the major commercially exploited stocks are showing signs of over exploitation. On the contrary, demand for seafood has been growing. The present scenario suggests that the current level of marine fish production from the exploited zone has to be sustained by closely monitoring the landings and the fishing effort and by strictly implementing the scientific management measures. The Government of Kerala imposed a trawl ban from 1988 onwards along the entire coastline of the state for a period of 45 days with effect from 15th of June in order to conserve the fishery wealth and thereby protect the interest of the persons engaged in fishing particularly those engaged in fishing using traditional fishing crafts as well as to regulate fishing on a scientific basis. Table 2.11 shows Indian seafood industry's installed capacity in the mid of second phase.

Exploitation of off-shore resources in the EEZ will have to be reconsidered in terms of not only the resources available, but also in terms of infrastructure. In order to avoid over-capitalisation and ensure a cautious growth of the infrastructure and investments, a rationalised approach will be essential in determining the number and size of fishing vessels, their resource specific gear as well as technology to be made available either indigenously or through foreign collaborations. The development of deep sea fishery industry is of concern to the entire marine fishery sector because it would have considerable impact on the management of near-shore fisheries, shore-based infrastructure utilisation and post-harvest activities, both for domestic marketing and export. The Deep Sea Fishing policy announced in 1991 permitted large industrial houses and multilateral companies to take up deep sea fishing and to enter into joint ventures. Increasing international competition, emergence of new suppliers, declining marine catch due to overfishing, insufficient development of aquaculture and inadequate exploitation of deep sea fishing resources are responsible for the decrease in the growth rates of marine products exports from Kerala.

Table 2.11: India's Seafood Industry As on 30.9.1991

Items	Nos	Installed Capacity/day Tonnes
Freezing Plants	249	2867.3
IQF facilities	45	228.3
Canning Plants	23	81.5
Fishmeal Plants	21	375.5
Frozen Storages	323	50680
Pre-Processing Centers	921	2176.78

Source: MPEDA, Government of India, Kochi.

The Kerala Fishermen's Welfare Fund Act 1985was amended, (Act 17of 1999) for incorporating the workers engaged in fishery related activities for his livelihood and to extend financial assistance to them from the Board. Allied workers includes beach workers, small scale fish distributors, fish cures, peeling workers and small scale processing plant workers who are not members or not eligible to get membership in any other Statutory Welfare Schemes. The present deep sea fishing has been undertaken by Matsyafed with four fishing vessels operating in Vishakpatanam. Matsyafed is also organizing programme to join with some foreign countries especially Japan and Singapore for joint venture to capture tuna at a larger scale. This would help to reduce the exploitation and competition in the in shore waters.

Third Phase: Ban Period and Quality Renovation: 1997 to 2004

India's foreign trade profile has changed drastically after liberalisation of Indian economy. Liberalization of import trade accelerates the growth of Indian imports. During 1990-91 the trade deficit amounted to Rs.10635 Crore. It was Rs. 65741 crore in the year 2003-2004, around 518.15 percent increase due to lower industrial and agricultural growth rates. At the third phase percent growth rate of trade deficit was around 173.06 percent. With the rapid pace of globalisation, removal of quantitative restrictions and a regime of free trade, increasing movement of live aquatic animals, threats of non-tariff trade barriers, imposition of sanitary and phyto-sanitary measures, etc, the importance of regional and international cooperation has assumed much more significance than what it was in the past. The third phase embarks upon the structural changes of seafood export processing industry that took place in India especially in Kerala.

Fish Production and Export Trend

Fish production in the country has been showing an increasing trend and has reached a record level of 5.65 million tonnes in 1999-2000. The estimated fish catch in the year 2000-01 was about 5.95 million tonnes and the production is likely to reach a level of 6.26 million tonnes by the end of the Ninth Five Year plan. However, the achievement of 6.26 million tonnes is much below the target of 7.04 million tonnes set for the Ninth plan at a growth rate of 5.64 percent per annum. Gujarat has emerged as the leading producer of marine fish during 1999-2000, followed by Kerala, Maharashtra and Tamil Nadu (Planning commission, 2001).Provisionally the current (1999-2000) annual fish production has been estimated at 5.65 mt - 2.83 mt from the marine sector against a potential of 3.9 mt and 2.82 mt from the inland sector against a potential of 4.5 mt. It is estimated that for every fishermen engaged in primary fishing activity about four others are getting additional employment by way of post-harvest operations, fish marketing and a host of other allied activities. However, the growth in marine fish production over the recent years has been rather sluggish at an average of 2.2 percent during the period 1991-1992 to 1999-2000 as compared to the inland fisheries at an average of 6.5 percent.

In the third phase of both inland and marine fish production in India showed fluctuating growth rates average annual growth rate is given in the Table 2.12 quantity of marine products from the level of 295 thousand tonnes in 1997-98 has decreased to 2941 thousand tonnes in 2003-04. While the inland fish production increased from 2483 thousand tonnes to 3458 thousand tonnes in 2003-04. Table 2.12 and figure 2.9 portray how marine fish production showed a downward trend and inland fish production upward trend in the third phase. Fish production trend in Kerala has stagnated despite enormous increases in fishing capacity. For certain species catch levels have declined. Mackerel production has dropped from 128,000 tonnes in 1996 to just 20,000 tonnes in 2001.

Table 2.12: Fish Production in India 1997- 2004 III Phase

Year	Fish Production (000 Tonnes)			Average Annual Growth Rate(Per cent)		
	Marine	Inland	Total	Marine	Inland	Total
1997-98	2950	2438	5388	-0.57	6.39	0.75
1998-99	2696	2602	5298	-8.61	6.73	-1.67
1999-00	2852	2823	5675	5.79	8.49	7.12
2000-01	2811	2845	5656	-1.44	0.78	-0.33
2001-02	2830	3126	5956	0.68	9.88	5.3
2002-03	2990	3210	6200	5.65	2.69	4.1
2003-04	2941	3458	6399	-1.64	7.73	3.21

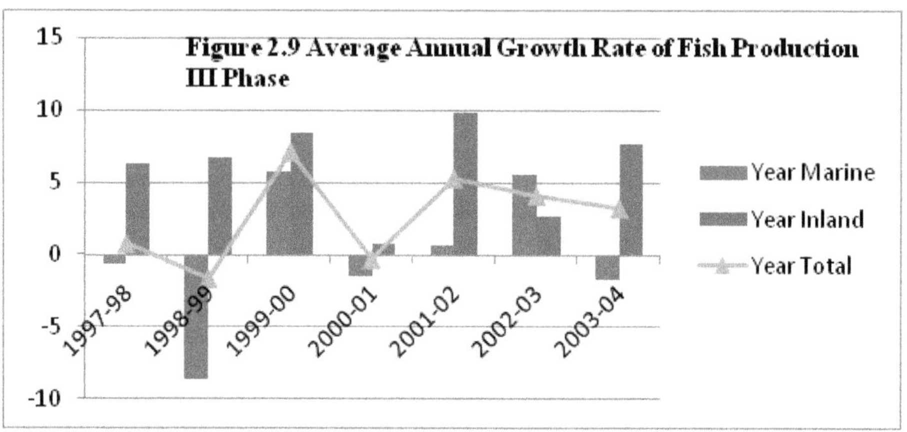

Source: Hand Book of Fisheries and Aquaculture, Indian Council of Agricultural Research, New Delhi 2014

Annual landings of shrimp declined from an average of 51,000 tonnes between 1985 to 1994 to an average of 46000 tonnes during 1995 to 2004. The production of sharks rays, ribbon fish, catfish, anchovies, goat fish, croakers, carangids and pomfrets show clear declining trends. Conversely, sardines, squid and seer fish show increasing trends. In India unlike China diversification in aquaculture has not taken place. India so far has culture technology of about 15 species whereas China has developed farming practices for 39 species (Planning commission, 2001).

In the value terms, world fishery product exports in 1998 stood at US $49 billion. The developing countries contributed to 48 percent of the total. The export of fish and fish products thus is very important for many maritime developing countries. When about 40 percent of global fish production enters international trade, only about 6 to 8 percent of forest products enter international trade (FAO 2000). The net foreign exchange earnings of developing countries in 1977 from fish and fish products stood at about US$16 billion, which according to FAO is more significant than the combined net export earnings from coffee, tea, rice and rubber. In the nineties, the export has increased with an annual growth rate of 10.41percent and 20.23percent in quantity and value respectively. Our marine products exports have shown consistently improved performance in overseas markets in view of the high quality.

Exports of marine products have maintained a steady growth. In Table 2.13 and Figure 2.10 indicates that our exports have crossed US$ 1.4 billion since 2001. Export of marine products has increased considerably to an all time high both in volume and value during 2002-03 with actual export of 467297 Metric Tonnes valued at Rs.6881 crore or US $ 1425 million. Figure 2.11 shows Kerala's share percentage in marine export products from India. During 1999-2000 Kerala's share in terms of value accounted for 22.41 percent and during 2001-2002 it declines in terms of value. However it increased to 18.04 during 2003-04.

Table 2.13: Growth of India Marine Products 1997-2004 III Phase

Year	Qty in Tonnes	Value in Rs. Crore	Average Unit value Realiza-tion (Rs. / Kg)	Average Exchange Rate US $	Value in US $ Million	Average Unit Value Realiza-tion US $ / Kg.	Growth rate % Qty	Growth rate %Rupee Value
1997-98	385818	4697.48	121.75	36.25	1295.86	3.36	2.01	13.98
1998-99	302934	4626.87	152.74	41.8	1106.91	3.65	-21.48	-1.5
1999-00	343031	5116.67	149.16	43.03	1189.09	3.47	13.24	10.59
2000-01	440473	6443.89	146.29	45.4975	1416.32	3.22	28.41	25.94
2001-02	424470	5957.05	140.34	47.5292	1253.35	2.95	-3.63	-7.56
2002-03	467297	6881.31	147.26	48.2933	1424.9	3.05	10.09	15.52
2003-04	412017	6091.95	147.86	45.7091	1330.76	3.23	-11.83	-11.47
2004-05	461329	6646.69	144.08	44.6683	1478.48	3.2	11.97	9.11

Source: MPEDA, Government of India, Kochi.

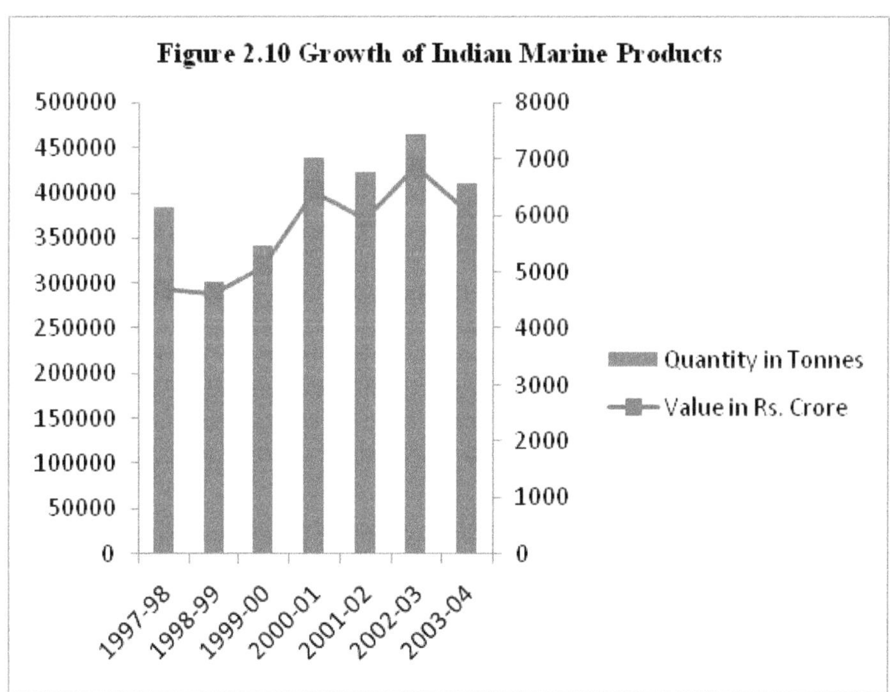

Source: MPEDA, Government of India, Kochi.

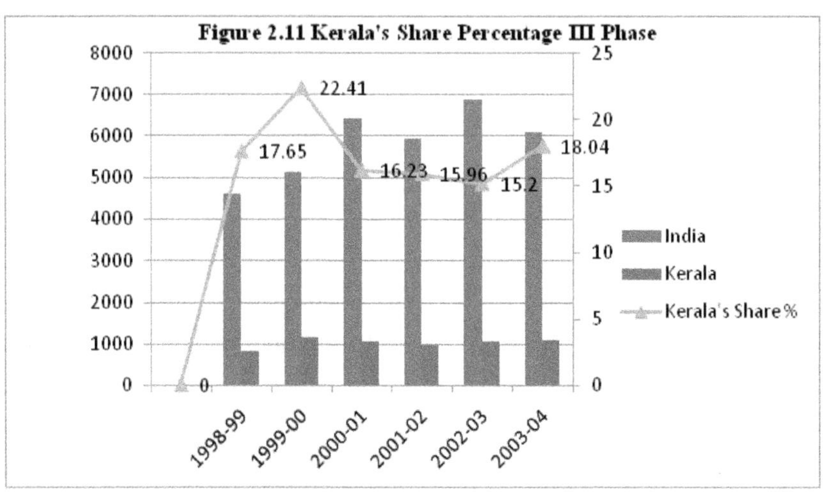

Source: CMFRI & MPEDA, Government of India, Kochi.

Destination Changes and Product Diversification

Globalisation essentially means integrating the economy of a country with rest of the world. Foreign trade provides an outlet to globalisation, which is being undertaken by most of the countries of the world today. India too has embarked on the policy of export-led growth strategy as the process of economic liberalisation initiated in the country since July 1991. Most seafood processing industries in India have surplus capacities and they should be encouraged to diversify production line, implement value addition and explore new market outlets. They must also be directed to enter domestic markets with processed quality fish. For a long time USA was the principal buyer for our frozen shrimp but after 1977, Japan emerged as the principal buyer of the product, followed by the West European countries. Japan retained its position till 2001-02 as the single largest buyer for our marine products accounting for about 31percent in the total export value. During the year 2002-03 and 2003-04 USA emerged as the single largest market for our marine products. Table 2.14 and Figure 2.12 shows the market-wise destination changes of seafood importing countries during the III Phase.

Table 2.14: Market Wise Export of Marine Products III Phase

Q: Quantity in M T, V: Value in Rs. Crore, $: US Dollar Million

Market		1997-98	1998-99	1999-00	2000-01	2001-02	2002-03	2003-04
JAPAN	Q:	70955	67277	66990	68983	64905	54916	50020
	V:	2326.09	2295.48	2272.78	2560.39	1820.69	1534.76	1163.69
	$:	641.68	547.55	527.68	562.75	383.07	317.17	253.86
USA	Q:	32914	34472	36645	41747	49041	61703	53153
	V:	583.75	617.32	775.35	1164.4	1421.38	2051.12	1682.06

Market		1997-98	1998-99	1999-00	2000-01	2001-02	2002-03	2003-04
		Q: Quantity in M T, V: Value in Rs. Crore, $: US Dollar Million						
	$:	161.03	147.61	180.11	255.93	299.05	424.51	365.84
EUROPEAN	Q:	34088	54081	66634	68827	82895	94541	96284
UNION	V:	405.89	682.61	912.03	1025.34	1150.07	1388.47	1470.99
	$:	111.97	163.49	211.71	225.37	241.97	287.84	319.95
CHINA	Q:	186537	87211	107136	182771	134767	170811	123738
	V:	816.93	483.02	544.7	827.42	597.23	762.48	676.46
	$:	225.36	115.52	126.33	181.86	125.66	158.23	151.6
SOUTH	Q:	30779	26917	38300	40748	52424	44097	50670
EAST ASIA	V:	308.56	267.95	360.19	462.97	538.75	642.38	545.77
	$:	85.12	64.1	83.64	101.76	113.35	133.15	119.13
MIDDLE	Q:	17310	16798	12460	17236	19159	19668	14711
EAST	V:	142.39	142.82	109.49	188.32	181.06	204.74	201.52
	$:	39.28	34.35	25.39	41.39	38.1	42.4	43.92
OTHERS	Q:	13236	16177	14867	20161	21279	21561	23441
	V:	113.87	137.67	142.13	215.05	247.87	297.36	351.46
	$:	31.41	32.86	33	47.26	52.15	61.6	76.46
Total	Q:	385818	302934	343031	440473	424470	467297	412017
	V:	4697.48	4626.86	5116.67	6443.89	5957.05	6881.31	6091.95
	$:	1295.86	1105.49	1187.87	1416.32	1253.35	1424.9	1330.76

Source. MPEDA, Government of India, Kochi.

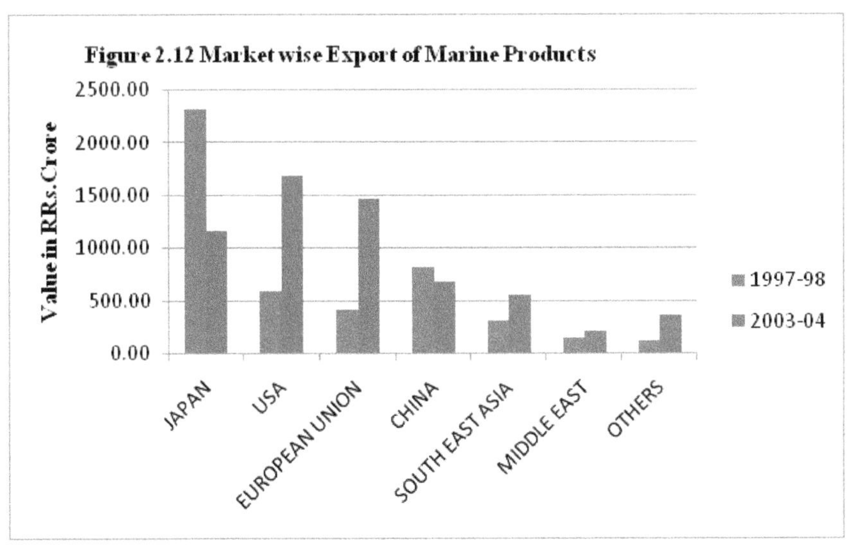

Source: MPEDA, Government of India, Kochi.

Nearly 55 categories of marine products are exported to countries in Southeast Asia, Europe and the US. The total quantity of export increased from 97,000 tonnes in 1987-88 to 3.85 lakh tonnes during 1997-98. In terms of value, it increased from Rs 530 crore to Rs 4,697.5 crore. Japan continued to be the largest market in terms of value with a share of 49.5 percent. Exports to China have grown by 500 percent between 1994-95 and 1997-98. The three-month ban on Indian seafood exports during 1997-98 by the European Union was one of the reasons for the significant growth in export to China. The export of marine products suffered a lot of problems like ban by EU on the plea that the products were infected with salmonella and cholera during August 1997.

Later on the ban was lifted. Recently, the USA has banned the marine products export from India to promote and facilitate the local fishermen of their country. However, India has maintained a steady marginal growth on the export of marine products to different countries. More than six million people earn their livelihood from this fishery industry. If all the available fishery resources are exploited in a suitable manner, a few more million employment opportunities can be created.

Globally 40 percent of fish production, in live weight equivalent, is traded. Export volume has increased three times towards the end of 2000 as compared to 1970s. Due to scientific shrimp farming, the export of frozen value added shrimp is continuing as the major foreign exchange earner among seafood products and the volume of frozen shrimp exported during 2003-2004 was 41, 2017 metric tonnes. Figure 2.13 depicts item wise marine products export from India during 2003-04.

Figure 2.13 Item wise Marine Products from India 2003-04

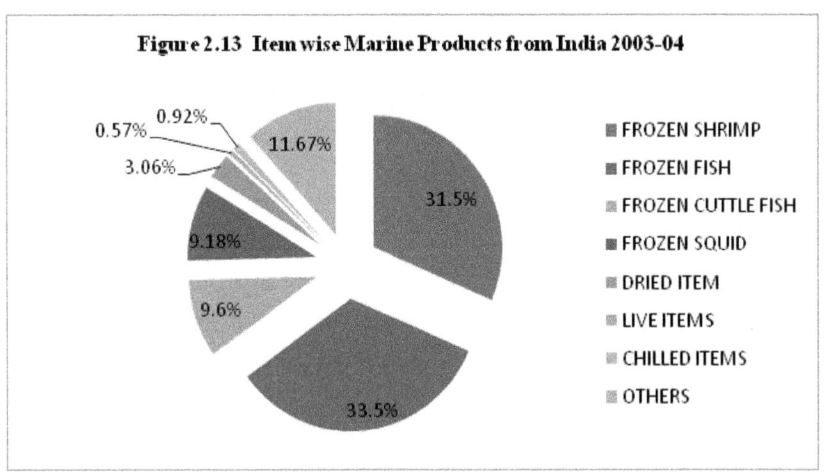

Source: MPEDA, Government of India, Kochi.

The period from 2001 onwards has been witnessing a shift in the marine fisheries scenario. During 2003 nearly 123,923 tonnes of frozen shrimps worth Rs. 3,936 crores were exported which formed 65.5 percent of total seafood export from the country. Employment of the primary capture fisheries and aquaculture production sectors in 1998 is estimated to have been about 36 million people, comprising about 15 million full-time, 13 million part-time and 8 million occasional workers. Employment in aquaculture (inland and marine) has been increasing and is now estimated to account for about 25 percent of the total. Marine capture fisheries account for about 60 percent and inland capture fisheries for the remaining 15 percent (FAO, 2000). In India, the total fishermen population as per the 1992 Census has been estimated at 6.7 million of which 0.74 million were reported as full-time fishermen.

Quality Revolution

In the third Phase, Stringent Quality Standards followed by EU, USA, Japan and the implementation of HACCP in seafood processing plants become mandatory. HACCP is a tool that properly used will help build confidence in the consumer's perception of seafood and the seafood industry. The 1990s have witnessed important international agreements and accords relating to the intentions of the international community to achieve sustainable fisheries and to which India has been a party. These agreements represent milestones in international efforts over many years and include UN Programme of action includes programme areas relating to coastal areas and the oceans; the 1992 International Conference on Responsible Fishing (held in Cancun, Mexico) and the 1993 and agreement to promote compliance with International Conservation and Management Measures by fishing vessels on the high seas.

In 1998, only 36 percent of the world fish underwent some form of processing. Improved post-harvest processing is seen as a way of developing the fishery industry without increasing harvests. In the marine sector, the available landing and berthing facilities is able to meet the requirements of about 25 percent of the available fishing fleet, leading to congestion and spoilage of fish. In the inland sector, marketing is again the weakest link. Value addition is also poor in India. Even though production in Japan is less than half of India, value realization per kg of Japan is US $ 4.37 compared to 1.11 $/kg of India. Reducing losses through proper handling and improved processing can increase fish production and also add value to the catch. Sanitary and phyto-sanitary measures are expected to play a major role in both domestic marketing and exports. Incorporation of HACCP and ISO 9000 into the post-harvest activities is a dire necessity and adequate attention would have to be paid in this area for providing better trade opportunities to the producer.

The EU had imposed a ban on the exports of all fishing products from India in 1997 marking reasons of health, safety and hygiene and revealed the initial trade impact of the SPS measures imposed by the EU which temporarily restricted market access of all the seafood firms exporting to the EU. The ban was lifted by the year end and firms must have compliance with EU health and safety standards. This was mainly due to buying of the European importers after lifting the ban on import of marine products from India in November 1997. By the end of 1990s, EU Team visited India again and approved 10 plants initially for exporting to the EU. By the end of 1990s it became imperative that Indian seafood maintained International quality standards. Today India has the largest number of EU approved processing plants.

The US FDA team visited India to assess the quality control measures adopted by processing plants and their conformance to US FDA regulations. Since 1999

India has been one of the leading users as well as victims of the anti-dumping measures. The period witnessed many uncertainties emerging in the marine fisheries sector consequent to the WTO framework and trade barriers. The Government of India released the National Marine Fishing Policy in the year 2004. It is intended for augmenting marine fish production, boost export, and increase in per capita consumption and mounting investment needs. Our country is trying to expand fishing activities in its 2.02 million km^2 offshore Exclusive Economic Zone (EEZ) where there may be scope for further growth. The USA insisted on DSP 121 certificate to confirm that the shrimp exported were caught without harming the habitat of turtles. This certificate was made mandatory with every consignment sent to the US. Turtle Excluder Device (TED) was given to all operating fishing vessels.

India has entered into trade agreements with several nations, which allow duty free export and imports, inclusive of fisheries items. Under the agreement signed between Government of India and Sri Lanka in 1999, the two countries decided to progressively reduce and eliminate obstacles in the way of bilateral trade through a free trade agreement. A number of fisheries items are allowed to be exported to India by Sri Lanka as duty free under this agreement. India also signed an agreement with SAFTA (South Asian Free Trade Area) in 2004, which comprises the SAARC nations. Preferential trade agreements too are signed between India and other South Asian nations such as Bhutan under SAPTA (South Asian Preferential Trade Area). In the case of fisheries items, Bangladesh and Maldives get preferential treatment, the former under SAPTA, and the latter in respect of certain items under SAFTA. Under the initiative of the Economic and Social Commission for Asia and the Pacific (ESCAP) an agreement was signed in 1975 for trade expansion through exchange of tariff concessions among seven developing member countries of the ESCAP region. These countries, namely, Bangladesh, India, Laos PDR, Republic of Korea, Sri Lanka, the Philippines and Thailand, agreed to a list of products for mutual tariff reduction. Certain fisheries items get concessions in terms of import tariffs under this Agreement. A preferential trade agreement was also signed by India with the Mercosur countries (Argentina, Brazil, Paraguay, and Uruguay) in 2004. While ornamental fish appear in the preferential list offered by India, no fisheries item appears in the preferential offer list of the other countries.

Technological & Infrastructural Challenges

From the 1990s, technological changes were accelerated with the globalisation policies. The proclamation of the policy on deep sea fishing in 1991 by the Central government that the Indian seas thrown open to the mechanised boats, trawlers and factory trawlers owned by Multi National /Supra National Companies for fishing. The exploitation of the fish resources becomes aggressive with these policies. For instance, a factory ship Oriental Angel anchored at the Kochi Port after deep sea fishing in the Kerala waters as part of this policy found to have fished a quantity equal to that would brought by 1500 trawling boats. The permission was given to 2630 such ships to operate in the Indian waters. The effect of the powerful trawling operations in the seabed badly affected the fish wealth of India. With globalisation, the fishery economy got a strong link with the international market and economy and the slight fluctuations were strong enough to make lasting repercussions.

European Union imposed sanctions on the Indian import in years. The sanction was due to the hygienic reasons. The detection of typhoid bacteria was also a major reason. This was a due to the urban wastage merging in the sea. This was coupled with the reduction of the cost of fish exported from India to China by 40percent. The sea became 'dry' due to the operation of the foreign trawlers, unscientific fishing etc. It was at this time of test that the price of diesel got increased and this was a serious problem for the fish workers and they observed a hartal in protest against the price rise. The price of the Kerosene was also hiked because of the reduction in the permit quota to the artisanal fishermen. Instead of the Rs.2.59 per liter, they have to pay Rs.12 to 16 in black market. The gravity of the issue was as large that many a small-scale boat owners (Individual and collective) suffered badly in this crisis. The reduction of the Kerosene permit limit from 1000 liters to 350 liters for the motorised boats also was a problem..

In 1999, the country had an estimated 1, 81,284 traditional fishing crafts; 44,578 motorised traditional crafts, 53,684 mechanised fishing boats and about 200 deep-sea vessels in operation. Fish processing in India is mainly for export. Open sun dried fish and fish meal are the only major exceptions. At present India has freezing units, cold stores, ice plants, canning units and fishmeal plants. Capacity of most of these processing and storage units is small when compared to the facilities in fish processing industry in technologically advanced countries. The total fish processing and storage facility in India is inadequate when compared to the potential for fish production and processing. Inland fisheries need low cost palletized feeds and special containers to transport fingerlings and fish. More rearing ponds are needed. Techniques to reduce seepage loss of water have to be introduced. Obsolete fishing gear needs replacement with better gear. Extensive network of refrigerated handling, transport, storage and retailing has to be put in place. Also, we have to make better use of fish waste and by-products.

Rapid technological advancement in fishing methods have resulted in higher usage of modern fishing equipment such as motor boats which are said to pollute and destroy natural aquatic environments. The guiding principle of fisheries management is maximum sustainable yield (MSY) which refers to the conjectural highest amount of fish that can be caught in each season without preventing fish stocks from regenerating. Fish is not part of the agricultural negotiations of the World Trade Organization (WTO) and continues to be treated as an industrial product in negotiations.

The Fourth Phase: US Anti-Dumping Duty on Indian Shrimp 2004 to 2014

Seafood export processing industry was affected by Tsunami in the fourth phase. It is among the most terrifying natural hazards known to man and have serious impact on human lives, social and economic sectors. Tsunami of December

2004 has dealt a severe blow to the coastal marine fishery sector causing huge loss of lives, fishing crafts and gears in the state especially in Thiruvananthapuram, Kollam, Allappuzha and Ernakulam districts. As per the official figures released by the Ministry of Home Affairs, the number of losses or missing of human lives was 176, population affected was 2470, villages affected were 187 and 11,832 dwellings were also affected in Kerala. Immediately after the Tsunami, fishing operations along the coastal regions came to a standstill. Few fishermen who ventured into fishing restricted their activities near to the shore.

Seafood industry has hard-hit by the tsunami, statistics indicated that the shrimp exports from India to the US was 30 percent lower in September 2005 compared to the year before, and that the number of Indian shrimp exporters have been cut in half due to the imposition of US antidumping duty on frozen shrimp imports from India. The Seafood Exporters Association of India (SEAI) has noted that the US market amounts to roughly 25 percent of the countries' total shrimp exports (ICTSD, 2005). The outbreak of the global financial crisis in 2008-09 led to the world trade in recession. India has been quite successful in diversifying its export markets from developed countries like the US and Europe to Asia and Africa, which has helped to a great extent in weathering the global crisis of 2008 and the recent global slowdown.

Fish Production and Export Trend

World trade of fish for human consumption is expected to reach 45 Metric Tonnes in live weight equivalent in 2023, up 20 percent from the base period but with the annual growth rate slowing from 2.7 percent in 2004 to 2013 to 1.7 percent in 2014-2023. This decline will be caused by increase in transportation costs, slower growth of fishery production and sustained domestic demand in some of the major exporting countries. Fish production in India increased from a level of 0.75 million tonnes in 1950-51 to 9.58 million tonnes in 2013-14 comprising 3.44 million tonnes from the marine and 6.14 million tonnes from inland resources. The fishery sector also accounts for 0.83 percent of total GDP and 4.75 percent of the Agriculture Sector's GDP at current prices for the year 2012-13. According to OECD–FAO (2013) projections, world fish production trend will achieve 181 million tonnes by 2022. The main source of growth is from aquaculture production because production in capture fisheries are stagnant since mid-1980s at about 85–95 million tonnes per annum due to depletion of fishery resources.

Recently, about 200 commercially valuable species are exploited from the marine fisheries resources from India. The marine production was estimated at 3.73 lakh tonnes in the year 1947-48 but declined to 2.27 million tonnes in the year 2005. Fish base has listed 2384 finfish species from Indian continent that includes 1704 marine, 762 freshwater, 202 endemic and 258 commercially exploited species (www.fishbase.org.).The marine fisheries sector in the country contributes about 81percent of the total fish production and is one of the major contributors to foreign exchange earnings through seafood export. It constitutes about 16percent of the total agriculture products export. Fisheries sector contributed Rs. 34,758 crores to the GDP during 2005-06 which was 1.2percent of the national GDP and to 5.3 percent of the Agricultural GDP. However the share of fisheries sector in the State Domestic product was estimated to be 1.44 percent in 2005-06. There has been a gradual shift in the production scenario from marine to inland fisheries in recent years (Economic Survey, 2007 and ICAR, 2006). The Gross State Domestic Product of the State has increased by about 97 percent during the period from 2005-06 to 2010-11 and the share of fisheries sector in the State Domestic Product has declined from 1.81 to 1.29 percent in the same period. The share of Primary Sector in GSDP has also declined from 18.05 to 14.07 percent in 2010-11.

Estimates of the fishery resources assessment shows that among the maritime states in India, Kerala occupies the second position in marine fish production. The total fish production in Kerala during 2012-13 was 6.8 lakh tonnes. The marine fishery resources of the state have almost attained the optimum level of production. At the national level more than 60 percent of the total fish production is contributed by the inland sector. Government has approved a master plan for increasing the inland fish production of the state from the current level of 75000 tonnes to 2 lakh tonnes over a period of 10 years. The Supply- demand gap may cause fish prices to rise in the future unless fish supply situation in the country is improved through appropriate measures (Sathiadhas et al., 2012). Marine Resources is almost reaching a stagnant level, measures to increase fish production through development and promotion of aquaculture and mariculture technologies is essential. This can be feasible with the help of public and private participation, implementing sustainable harvesting strategies in the marine sector and efficient domestic value chain management. These are required for reducing the price level and meeting the fish consumption in the future. Figure 2.14 depicts the marine fish landings in India and Kerala during the IV phase.

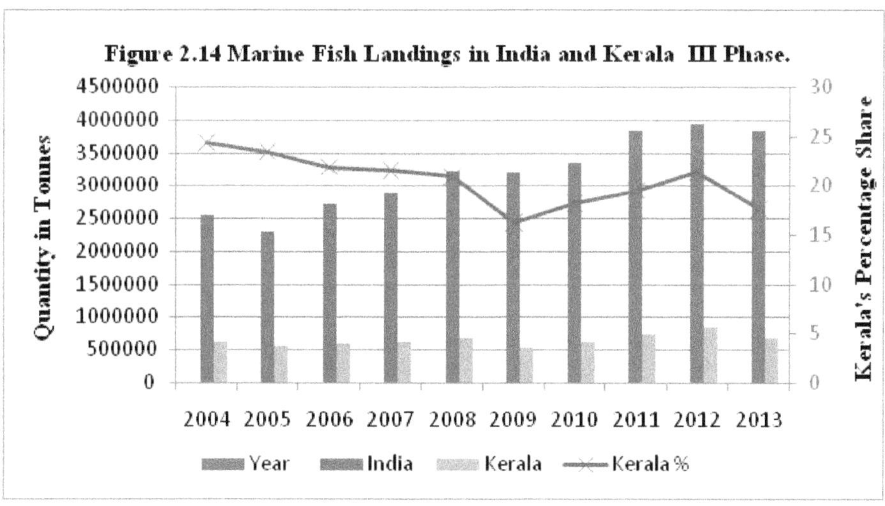

Source: CMFRI, Government of India, Kochi

In Kerala marine fish landings remained stagnant at 6 lakh tonnes during 2000-2010. The value of fish catch at landing centre level increased from Rs. 2,438 crores to Rs. 3,803 crores and at retail level from Rs 4272 crores to Rs. 5520 crores during 2000-2010. Even though the catch slightly increased during the years 2007, 2008 and 2010 there was no increase in the average fish price as the fish catch was dominated by the low value clupeids. In Kerala, the catch of penaeid prawns declined from 56462 tonnes in 2000 to 35652 tonnes in 2010 whereas the catch of oil sardines reached 259342 tonnes in 2010.Nearly one third of the fish catch is contributed by oil sardines which resulted in the lower average prices in the state. Very low landings of oil sardines and comparatively high landings of prawns might have caused the high average prices in 2009.

The National Fishery Development Board was inaugurated in 2006.During 2006-07 the catch of Ribbon fish was 18000 tonnes. Unfortunately the share of these high value varieties in the total marine fish catch has been remaining stagnant. The annual potential of prawn is estimated at 64482 tonnes in 2006-07.

Table 2.15 GROWTH IN EXPORT OF INDIAN MARINE PRODUCTS IV Phase

Year	Qty in Tonnes	Value in Rs. Crore	Average Unit value Realization (Rs. / Kg)	Average Unit value Realization US $ / Kg.	Growth rate % Qty	Growth rate %Rupee Value	Value in US $ Million
2004-05	461329	6646.69	144.08	3.2	11.97	9.11	1478.48
2005-06	512164	7245.3	141.46	3.21	11.02	9.05	1644.21
2006-07	612641	8363.53	136.52	3.02	19.62	15.43	1852.93
2007-08	541701	7620.92	140.68	3.51	-11.58	-8.88	1899.09
2008-09	602835	8607.94	145.79	3.17	11.29	12.95	1908.63
2009-10	678436	10048.53	148.11	3.14	12.54	16.74	2132.84
2010-11	813091	12901.47	158.67	3.51	19.85	28.39	2856.92
2011-12	862021	16597.23	192.54	4.07	6.02	28.65	3508.45
2012-13	928215	18856.26	203.14	3.78	7.68	13.61	3511.67
2013-14	983756	30213.26	307.12	5.09	5.98	60.23	5007.7

Source: MPEDA, Government of India, Kochi

The export of marine products has steadily grown over the years from a mere Rs.3.92 crore in 1961-62 to Rs. 30213.26 crore in 2013-14 and is shown in Table 2.15 and Figure 2.15. Exports of marine products from India during the financial year 2009-10 touched the $2.13 billion mark, growing 12.54 percent in quantity, and 16.74 percent in rupee value. The marine products industry in India had crossed the $5 billion mark in 2013-14.

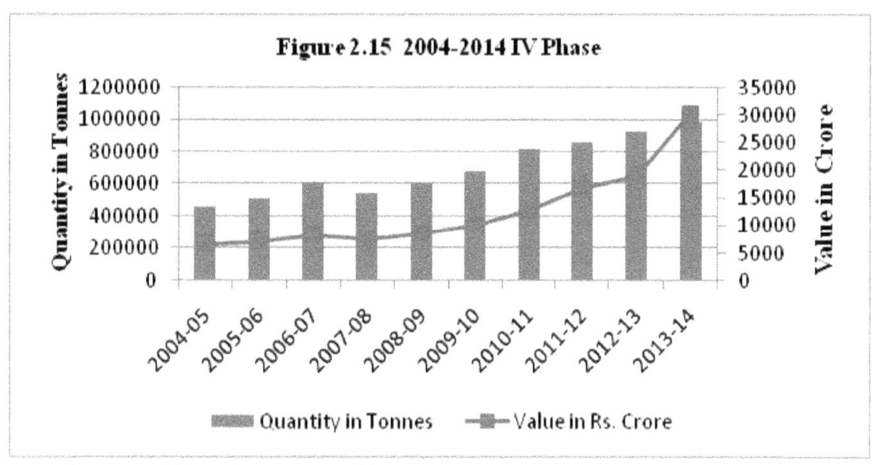

Source: MPEDA, Government of India, Kochi.

The Marine Products Export Development Authority pointed out that this achievement was made despite the after-effects of the recession in the international market, a stronger rupee vis-à-vis the euro and the dollar and the impact of the collapse of the economies of Greece, Spain and Portugal, causing the depreciation of euro against the dollar. Marine products exports from India during 2012-13, both in quantity and value, was far larger than the average quantity exported and value earned respectively during the 11th Five Year Plan period. The average quantity exported during 11th Plan was 699617 Metric Tonnes and the export of 2012-13 was 928215 Metric Tonnes. As such the average value of marine products exports during the 11th Plan was 1115522 Metric Tonnes and the value of 2012-13 was 1885626 lakh. In the case of Kerala also the same trend can be seen. The average contribution of Kerala to total quantity of marine exports during 11th Plan was 117744 Metric Tonnes with an average value of 193270 lakh whereas Kerala share in total marine exports during 2012-13 is 166399 Metric Tonnes in quantity and 343585 lakh in value terms (Economic Review, 2013).The Centre has set a target of doubling the country's marine exports to USD 7 billion by 2015 from USD 3.5 billion at present and is in the process of charting better schemes to ramp up public-private partnerships.

It also aims at enhancing and developing the aquaculture for the economic and development of the state and ensuring food and livelihood security to the people of Kerala. The current level of Inland fish production is to the quantum of about 1.21 lakh tonnes. The National Agriculture Policy, which aims to attain a growth rate in excess of four percent per annum in the agriculture sector, stresses the importance of food and nutritional security issues and the importance of animal husbandry and fisheries sectors in generating wealth and employment. India is one of the major fish producing countries in the world with third position in fisheries and second in inland fish production. During 2010-11, India's total fish production was 6.82 million tonnes of which 3.07 million tonnes were from marine sector and 3.75 million tonnes was from inland sector. The fisheries sector has contributed 1.1 percent of the National GDP and 4.7 percent of the GDP from agriculture sector.

The marine fish production in Kerala tended to oscillate and the inland fish production showed a sign of improvement from 2007-2008. During 2010-11, the marine fish production has decreased to 5.6 lakh tonnes from 5.70 lakh tonnes in 2009-10. Inland production sustained an increasing trend. District wise marine fish production showed that Alappuzha contributed the highest (23.08 percent) followed by Kollam (20.35 percent) and Kozhikode (14.95 percent). During 2010-11 the share of inland fish production to the total fish production of the state was 17.78 percent. The details of fish production in Kerala during the IV phase given in Table No. 2.16

Table 2.16: Fish Production in Kerala during 2004-05 to 2013-14 (P) Quantity in 000 tonnes

Year	Inland Fish Production	Marine Fish Production	Total Fish Production
2004-05	76.45	601.86	678.31
2005-06	77.98	558.91	636.89
2006-07	79.57	598.06	677.63
2007-08	81.04	586.29	667.33
2008-09	102.84	583.15	685.99
2009-10	128.84	570.01	698.86
2010-11	121.21	560.4	681.61
2011-12	140.03	553.18	693.21
2012-13	149.1	530.64	679.74
2013-14	186.34	522.31	708.65

Source: Directorate of Fisheries. Government of Kerala, Thiruvananthapuram

Kerala was the pioneer in marine product exports and its share is decreasing year after year. The share of Kerala in the total export of marine products was more than 72 percent in terms of value and 50 percent in terms of quantity during the sixties. By 1980-81 the share dropped to 38.56 percent in quantity and 40.59 percent in value. In 1990-91 volume and value of marine exports from Kerala marked 36.58 and 32.30 percent respectively. In the year 2001-02 the share has further declined to 7.14 percent in quantity and 15.96 percent in value. In 2004-05 the share has increased to 18.9 percent in quantity and 17.4 percent in value. In the year 2013-14 the share has further declined to 16.84 in quantity and15.58 in value and is shown in the Table No. 2.17. This is mainly due to the growth of culture fisheries in states like Andhra Pradesh, Orissa, and West Bengal etc. Even though the percentage share of Kerala is coming down, the volume and value of export from Kerala is generally on an increasing trend, except in the year 2013-14. The stagnation in capture fisheries has created raw material scarcity and affected the export growth from Kerala. Kerala contributes more than 40 percent in volume and about 50 percent in value of the total cuttlefish exported from India, 40 percent in volume and 43 percent in value of frozen squid, 25 percent in volume and 35 percent in value of frozen shrimp, 12 percent in volume and percent in value of frozen fish.

Table 2.17: Exports of Marine Products from Kerala 2004-2014 -IV Phase

Year	Quantity in Metric Tonnes Values in Rs Crores			
	Kerala		Share of Kerala percentage	
	Quantity	Value	Quantity %	Value %
2004-05	87378	1158	18.9	17.4
2005-06	97311	1258	18.9	17.36
2006-07	108616	1524.12	17.73	18.22
2007-08	100318	1430.94	18.52	18.78
2008-09	100780	1572.18	16.72	18.26
2009-10	107293	1670.02	15.81	16.62
2010-11	124615	2002.10	15.33	15.52
2011-12	155714	2988.33	18.06	18
2012-13	166399	3435.85	17.93	18.22
2013-14	165698	4706.36	16.84	15.58

Source: MPEDA, Government of India, Kochi.

The marine products export from the State during 2010-11 was valued at Rs.2002.10 crore constituting 15.33 percent in terms of volume and 15.52 percent in term of value in the Indian marine product export. The state's share in all India exports has been declining in recent years. The share declined from 18.9 percent in quantity terms in 2004-05 to 16.84 percent in 2013-14 and the share in value decreased to 17.4 percent from 15.58 percent.

Destination Changes and Product Diversification

Kerala is a coastal state and is bordered on the west by the marine flora and fauna rich Arabian Sea. Marine fish landings of India in 2010 was 3.07 million tonnes which recorded a decline of about 1.31 lakh tones compared to the estimate for 2009. Among the states, Tamilnadu was the highest contributor of marine fish followed by Kerala. During 2010-11, 5.6 lakh tonnes of marine fish were landed in Kerala showing a decrease of 0.10 lakh tonnes (1.75 percent) over the previous year. The fish catches from the Kerala coast include more than 300 different species; the commercially important number is about 40 only. The high value species among the fish catches are still few and the prominent among them are Seer fish, Prawn, Ribbon fish and Mackerel. During 2010-11 the catch of Ribbon fish was 15196 metric tonnes and penaeid prawn was 47620 metric tonnes. The quality of these high value species in the total catch ultimately decides the income of the fishermen. Oil sardine accounted for the major share of landings (27 percent), heavy landing of juvenile oil sardine in ring seine was also recorded. The catch of oil sardine was 151839 Metric Tonnes.

In the year 2004-05 European Union collectively became the largest buyer from India. During 2004 to 2006 European Union was the chief importer of seafood from India. India is exporting raw material to China, Thailand and Vietnam for value addition and re-exports to Japan, the EU and the US. Thus the scope for value-addition in marine exports sector should be further explored. The European Union continued to be the largest market with a share of 26 percent in dollar realisation. However, there was a marginal decline of 1 percent in the quantity exported to these countries. The U.S. regained the second place with a share of 16 percent, followed by South East Asia, also with a share of 16 percent, China with a share of 15 percent, Japan 14 percent, West Asia 5 percent and other countries 8 percent. The exports to the U.S. registered a growth of 104 percent in dollar realisation and 47 percent in terms of quantity. The exports to Japan also registered a positive growth of 11 percent in quantity and 36 percent in dollar terms. The exports of all items to Japan, except those frozen, showed an increasing trend. The South-east Asian countries had registered a positive growth of 44 percent in quantity and 38 percent in dollar terms. The exports to China showed only an increase of 5 percent in quantity and 9 percent in dollar terms.

According to the Economic Survey 2012-2013, growth in exports can only be achieved with greater diversification of products. India has been fairly successful in diversifying its export markets from developed countries like the US and Europe to Asia and Africa, which has helped to a great extent in weathering the global crisis of 2008 and the recent global slowdown (Table No. 2.18).

Table 2.18: Region-Wise Share of India's Exports

Sl.No.	Country	2000-01	2005-06	2011-12	2012-13 (Apr- Nov)
1	Europe	25.9	24.2	19	18.7
2	Africa	5.3	6.8	8.1	9.6
3	America	24.7	20.7	16.4	19.5
4	Asia	37.4	46.9	50	50.4
5	CIS & Baltics	2.3	1.2	1	1.3

Source: Computed from The Directorate General of Commercial Intelligence & Statistics, Government of India, Kolkata.

Over 55 varieties of marine products are exported to different countries in South East Asia, Europe, China, Japan and USA. The shrimp exports increased by 13 percent in quantity, 35 percent in rupee value and 41 percent in dollar value. The unit value realisation went up by 24 percent. The exports of value-added products showed a growth of about 2 percent in quantity and 12 percent in dollar realisation. The other major items of export are Frozen Fin fish, frozen cuttlefish, etc. Tuna fish is the third most traded fish internationally. Tuna fish exports are targeted to reach 400 million dollars by 2010.Andaman and Nicobar Island holds 25 -30 percent of the potential. The production of sharks, rays, ribbon fish, catfish, anchovies, goat fish, croakers, carangids and pomfrets show clear declining trends. Conversely, sardines, squid, and seer fish show an increase in trend.(SIFFS2007).

Table 2.19: Market Wise Exports of Marine Products (2004-2014)

Market	2004 -2005	2005 -2006	2006 -2007	2007 -2008	2008 -2009	2009 -2010	2010 -2011	2011 -2012	2012 -2013	2013-2014
JAPAN	57832	59785	67437	67373	57271	62690	70714	85800	76648	71484
	1202.45	1155.97	1353.38	1227.59	1234.01	1289.58	1683.39	2140.67	1999.6	2463.8
	266.96	262.79	299.2	305.49	278.61	278.56	373	456.35	372.57	410.95
USA	50045	55817	43758	36612	36877	33444	50095	68354	92447	110880
	1556.09	1639.24	1347.8	1016.94	1021.55	1012.52	1990.26	2977.53	4026.5	7744.7
	345.52	372.62	297.08	253.05	227.29	213.52	438.49	637.53	747.45	1286
EUROPEAN	117742	136842	149773	149381	155161	164800	170963	154221	158357	174686
UNION	1819.28	2134.25	2760.32	2664.24	2854.07	3013.33	3459.4	3810.44	4176.4	6129.7
	405.4	484.02	610.95	663.18	635.34	637.4	765.15	805.38	777.41	1013.3
CHINA	124826	137076	203513	139792	147312	144290	159147	84515	87776	75783
	693.25	849.45	1156.96	1009.59	1296.39	1790.89	1977.81	1259.23	1444.9	1766.7
	154.1	191.99	259.06	252.9	281.9	379.7	440.1	263.3	269.47	293.12
SOUTH	63842	60140	67650	63818	88953	149353	233964	343962	340944	380061
EAST ASIA	628.83	585.85	616.7	573.97	873.09	1479.55	2114.48	4193.27	4357.3	8046.6
	139.77	132.7	136.43	143.5	191.08	314.85	469.36	880.09	811.8	1321
MIDDLE	16624	22270	23585	25752	27177	34907	43983	38155	41419	58040
EAST	244.42	307.65	371.06	393.96	475.72	553.55	670.35	894.38	1113.3	1599.4
	54.7	69.64	82.47	98.05	105.2	117.05	148.31	186.85	209.26	272.65
OTHERS	30418	40234	56924	58972	90083	88953	84225	87014	130623	112822
	502.37	572.9	757.3	734.62	853.11	909.11	1005.77	1321.72	1738.3	2462.4
	112.03	130.44	167.75	182.93	189.22	191.77	222.5	278.94	323.71	410.71
Total	461329	512164	612641	541701	602835	678436	813091	862021	928215	983756
	6646.69	7245.3	8363.53	7620.92	8607.94	10048.5	12901.5	16597.2	18856	30213
	1478.48	1644.21	1852.93	1899.09	1908.63	2132.84	2856.92	3508.45	3511.7	5007.7

Source: MPEDA, Government of India, Kochi.

The details of major markets for Indian marine products are given in the Table 2.19 and Figure 2.16 depicts the destination changes due to global recession and international trade barriers in the seafood industry in India. South East Asia continued to be the largest buyer of Indian marine products with a share of 26.38 percent in terms of US $ value realization. USA is the second largest market with a share of 25.68 percent followed by European Union, 20.24 percent, Japan 8.21 percent, other countries 8.20 percent, China 5.85 percent and Middle East 5.45 percent.

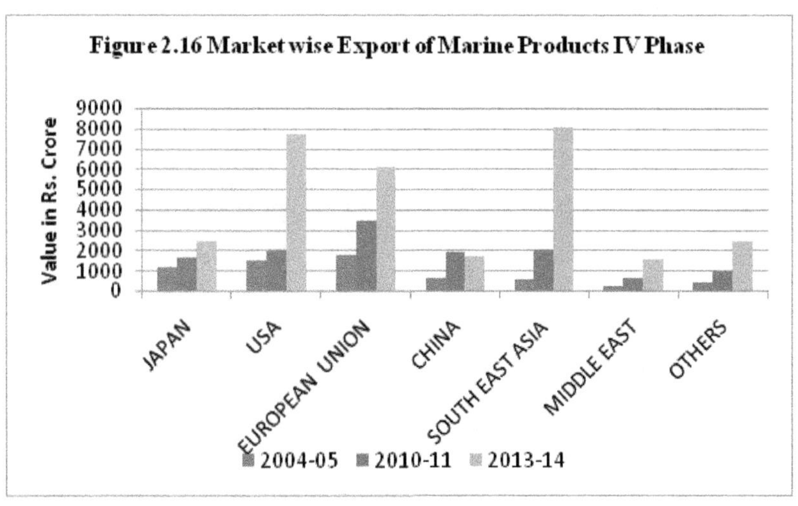

Source: MPEDA, Government of India, Kochi.

The exports to South East Asian Countries has shown a positive growth an increase of 11.47 percent, 84.67 percent and 62.72 percent in terms of Quantity, Rupee value and US dollar terms respectively. Exports to US had registered a tremendous growth of 19.94 percent in quantity and 72.06 percent in US$ realization and is mainly attributed to the export of Frozen Shrimp which showed a growth of about 34.81 percent in volume and 92.4 percent in US$ terms. Exports of Vannamei shrimp showed a tremendous growth to US market and the increase is 59.63 percent in quantity and 135.71 percent in US $ realization.

Export to Japan registered an increase in terms of US $ by 10.30 percent. Export of frozen shrimp increased by 7.38 percent in quantity terms and 28.23 percent in dollar terms. Export to Middle East countries has shown good growth of 40.13 percent, 43.65 percent and 30.29 percent in terms of Quantity, Value and Dollar terms respectively.

The details of major items of exports are given in the Figure 2.17 and show the percentage share quantitatively. Frozen shrimp continued to be the major export value item accounting for a share of 64.12 percent of the total US $ earnings. There was an all time high growth in unit value realization of frozen shrimp at 35.05 percent. The overall export of shrimp during 2013-14 was to the tune of 3, 01,435 Metric Tonnes worth US $ 3210.94 million. USA is the largest market (95,927 Metric Tonnes) for frozen shrimps exports in quantity terms followed by European Union

(73,487 Metric Tonnes), South East Asia (52,533 Metric Tonnes) and Japan (28,719 Metric Tonnes).The contribution of cultured shrimp to the total shrimp export is 73.31 percent in terms of US $. The export of cultured shrimp has shown tremendous growth of 36.71 percent in quantity and 92.29 percent in dollar terms. The export of Vannamei shrimp has shown tremendous growth to 1, 75,071 Metric Tonnes from 91,171 Metric Tonnes and US $ 1,994.27 million from 731.01 million when compared to 2012-13. Out of the total export 44.59 percent of total Vannamei shrimp was exported to USA followed by 17.07 percent to EU, 16.54 percent to South East Asian countries and 4.01 percent to Japan in terms US $. Export of Black Tiger shrimp reduced from US $521.33 million to 435.79 million and 61,177 Metric Tonnes to 34,133 Metric Tonnes in Quantity terms when compared to last year.

Fish, has retained its position as the principal export item in quantity terms and the second largest export item in value terms, accounting for a share of about 32.97 percent in quantity and 14.15 percent in US $ earnings. Unit value realization of fish also increased by 21.65 percent .Export of Frozen Squid has shown a growth of 15.98 percent , 25.68 percent and 10.78 percent in terms of Quantity, Rupee Value in and US $ terms. However it has shown a decrease in unit value realization of 4.48 percent. Frozen Cuttlefish recorded a growth of 8.34 percent in quantity. Dried items have shown a positive growth and the increase is 21.72 percent in terms of rupee value and in dollar terms by 9.86 percent. Live items exports shown a growth by 16.17 percent, 42.43 percent and 26.81 percent in quantity, rupee value and US $ realization respectively when compared to the previous year.

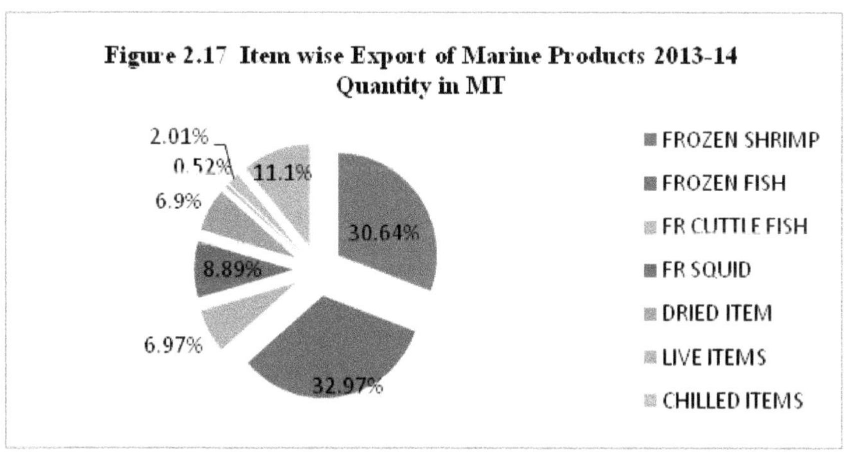

Figure 2.17 Item wise Export of Marine Products 2013-14 Quantity in MT

Source: MPEDA, Government of India, Kochi.

About 80 percent of the catch is channelised through domestic marketing system and the rest for exports. Exports improved from Vizag, Chennai, Krishnapatnam, Tuticorin and Mangalore compared to the corresponding period during the last year. Pipavav port is the major port in terms of quantity, 25.27 percent and Vizag is the major port in terms of dollar value, 22.59 percent. MPEDA envisages a target of USD 6.0 Billion for the year 2014-15. Increased production of Litopenaeus.

Vannamei shrimp, Quality control measures and increase in infrastructure facilities for production of value added items are expected to help in achieving this target.

Anti-dumping duty on Indian Shrimp Export to USA

The US anti-dumping duty on frozen shrimp imports from India was imposed from August 4, 2004. The average duty imposed on Indian companies was 10.17 percent and in the first Administrative Review (AR) this was cut to 7.22 percent. It was further reduced to 1.69 percent in the second AR and to 0.79 percent in the third. In the fifth AR, this was raised to 1.69 percent and it has been further enhanced to 2.51 percent. Between July, 2008 and May, 2009 the value of world trade declined by 37 percent, of which 16 percent was due to the fall in prices, a result of the financial and economic crisis. The WTO estimates projected that global trade is likely to decline by 9 percent in volume terms and the IMF estimates projected a decline of over 11 percent for 2009. Though India has not been affected to the same extent as other economies of the world during this phase, yet the declining trend in the growth rate of our exports and imports, have started in the second and third quarter of the year 2008-09 respectively. Even as export growth has surpassed the pre-crisis level in 2010-2011, the adverse impact of the global recession seems to be showing only now, with exports falling drastically in 2012-2013.

Seafood exports to the U.S. had been going down since the past three or four years mainly due to the anti-dumping laws of that country. As a result of this the European Union had emerged as the biggest seafood buyer from India both in terms of quantity and value that stands at 28 percent and 34 percent respectively. Quantity-wise, only seven percent of the seafood exports from India went to the U.S. and in terms of value it was 13 percent of the seafood export earnings.

In order to overcome the loss from the U.S, the seafood industry here had many potential markets that had not been tapped. The reason for this was that stock produced was only sufficient enough to meet the demands from traditional markets. The potential markets that had not been tapped included Russia, South Africa and Australia. The Indian seafood export industry was capable enough to remain very healthy even if exports to the U.S. came down to nil if the untapped markets were properly exploited.

The major barriers in the seafood export market in the fourth phase have been in the form of sanitary and phyto-sanitary measures imposed by the developed countries and the anti-dumping duty of frozen shrimp imposed by US. This adversely affected the prospects of Kerala Seafood Export processing industry. The strict enforcement of quality standards by the US and the EU led to hasty implementation of Codex Standards and HACCP regulations which generate additional cost of both fixed and variable costs which ranged between10 to 20 percent of the total cost (MPEDA, 2003).US Anti-Dumping duty imposition on Indian Frozen Shrimp (U S anti-dumping petition against six countries- Brazil, China, Ecuador, India, Thailand and Vietnam). Shrimp exports to USA dropped by 23 percent in Dollar earnings during 2006-07.After the Anti-dumping duties came into effect, the number of Indian exporters to the United States declined in a significant way from 280 in 2005 to just 68 in 2009. At present 192 exporters do shrimp export business with USA.

Technological & Infrastructural Facilities

Fish and Fishery trade has been growing very rapidly in recent decades. Exports played a significant role in the socioeconomic scenario of Kerala. During 2004, the marine fish catch was 2.81 million tonnes, of which 63 percent was taken from the west coast and the rest from the east coast. The fishing units consist of 208000 traditional craft, 55000 traditional motorized craft, 1250 mechanized boats and about 100 deep-sea fishing vessels. Table 2.20 shows the region wise processing plants with capacity both in India and Kerala. There are about 185 wholesale markets and 2518 retail market in Kerala (Department of Fisheries, Government of Kerala). The major fishing harbours in the state like Neendakara, Cochin, Munambam and Beypore are important primary trading centers also. The main supply side intermediaries in Kerala are auctioneers and commission agents, retail traders and agents of exporters.

Table 2.20 Region-wise Processing Plants With Capacity (Category : EU/Non EU)

Year	Kerala		India	
	No	Capacity	No	Capacity
2005	81	2370.83	261	10259.93
2006	85	2542.73	278	11096.33
2007	90	2770.73	295	12004.53
2008	95	2857.63	328	13540.93
2009	97	2882.63	349	14449.69
2010	102	2976.91	374	15451.15
2011	75	3021.31	391	16503.8
2012	111	3279.81	423	17477.04
2013	112	3309.91	447	18520.48

Source: National Informatics Centre Lakshadweep Office,Willingdon Island,Kochi

A community pre-processing centre, established by the Kerala Region of SEAI has poised for operation at Ambalapuzha in Alappuzha District of Kerala on 29th November 2010. The pre-processing plant is expected to eliminate unhygienic peeling practices which are now widely prevalent and also to upgrade the hygienic standards and sanitary conditions so as to bring them at par with international norms. The pre-processing plant consisting of 10 pre-processing units is capable of accommodating about 500 workers at a time. The facility include two tube ice machine of 10 ton capacity each, two generator sets and two chill rooms - one for storing raw material and other for storing finished products. Apart from these, a fully fledged laboratory is provided to test the raw material for microbiological parameters. Four bore wells and a water treatment plant for supply of potable water for all the units and an effluent treatment plant with sufficient capacity to treat the effluent before disposing to sea are also available at the centre. This project is first of its kind in India and the area is identified due to the availability of skilled workers and proximity to raw material sources. On account of the significance of the area in fish processing, the establishment of a Community Pre-processing Centre was thought of in the region. SEAI contributed sufficient land and manpower support while infrastructure, machinery and other or grant-in-aid was from ASIDE fund for this novel project.

Fishing efforts are largely confined to the inshore waters through artisanal, traditional, and mechanised sectors. About 90 percent of the present production from the marine sector is from within a depth range of up to 50 to 70 meters and remaining 10 percent from depths extending up to 200 meters. While 93 percent of the production is contributed by artisanal, mechanised and motorised sector, the remaining 7 percent is contributed by deep sea fishing fleets confining their operation mainly to the shrimp grounds in the upper East Coast. The Govt. of India released the National Marine Fishing Policy in the Year 2004. The Kerala State Fishing Boat Operators Association has protested against the proposed Marine Fisheries (Regulation and Management) Act 2009 and said that it contains hidden ideas of granting permission to foreign vessels to conduct industrial fishing in Indian seawaters. Deep sea fishing operations in the EEZ come under the purview of the Union Government. State Government has the right to decide on the type of vessels to be banned in its vessels in its territorial waters. Table 2.21 depicts the actual picture of built up capacity of maritime states of the Indian Seafood Industry.

Table 2.21: Built up Capacity of the Indian seafood Industry-2014

Name of the State	No. Of Exporters	No. of Process Plants	Freezing Capacity (Tonnes.p/d)	No. of Cold Storages	Storage Capacity	No. of Fishing Vessels
Kerala	287	124	1585.77	169	23086.5	2963
Tamil Nadu	286	48	524.55	67	5900	1562
Karnataka	43	14	186.4	26	3540	3226
Andhra Pradesh	95	52	779.5	53	7200	717
Goa	9	7	104	9	1275	420
Gujarat	64	55	2216.03	57	22925	426
Orissa	30	21	220	20	2460	414
Maharashtra	268	41	1327.11	39	19372	2932
West Bengal	99	37	340	30	3500	0
Delhi (UT)	92	--	0	1	15	0

Source: MPEDA, Government of India, Kochi.

Marine fisheries sector throughout the world is passing through a critical phase due to the present rate of biodiversity loss and fishing practices. Indian marine fisheries is also passing through a crisis due to its over capacity and open access nature. Marine fish production in India has showed a declining trend over the previous year. Estimates of the fishery resources assessment shows that among the maritime states in India, Kerala occupies the 2nd position in marine fish production. Frozen shrimp is continued to be the single largest item of export in terms of value. It accounts 44.32 percent of the total export earnings.

Potential for Dry Fish industry

There is a vast scope for the growth of dry fish processing in India, if only hygienic, modern and scientific methods of handling, processing, drying and preservation are adopted by the traditional Indian fishermen. Today, unfortunately, big players are fighting shy of the dry fish market, because of which there is a paucity of fresh investment into this sector, leaving the traditional fishermen to continue to follow the age-old traditional methods of unscientific dry fish processing and drying.

Economics of Family Farming

General Assembly of the United Nation in its 66th session adopted 2014 as the International Year of Family Farming. It includes all family based agriculture activities. It aims at repositioning family at the centre of agricultural, environmental and social policies in the national agenda by identifying gaps and opportunities to promote a shift towards a more equal and balanced development. It helps to raise the profile of family farming and smallholder family by focusing world attention on its significant role in the fight for eradication of hunger poverty, providing food security and nutrition, improving, livelihoods, managing natural resources, protecting the environment, and achieving sustainable development particularly in rural areas.

Value Addition

The food habits of the people all over the world are changing rapidly and India is gearing up to produce and supply value added products in convenience packs by adopting the latest technologies and by tapping the unexploited and under exploited fishery resources. Value addition has been considered as the main thrust area for seafood export processing units from Kerala. These units will be encouraged for value added production by expanding their capacity and diversifying their activities through foreign collaboration, investments, tie ups in marketing of value added products and importing raw materials for further processing and re-export in the form of value added products.

Blue Revolution

Blue Revolution means the adoption of a package programme to increase the production of fish and marine products. Blue Revolution in India was started in1970 during the Fifth Five Year Plan when the Central Government sponsored the Fish Farmers Development Agency (FFDA).Table No.2.22 shown FFDA achievement details from the Department of fisheries from Kerala. This programme was initiated to encourage the aquaculture, upgrade technology and encourage involvement of private sector for activities such as quality seed, feed and other inputs and creation of suitable infrastructure for storage, transport, marketing and credit. Subsequently Brackish Water Fish Farmers Development Agency was set up to develop Aquaculture. The Blue Revolution has brought improvement in aquaculture by adopting new techniques of fish breeding, fish rearing, fish marketing and fish export. Under the Blue Revolution Programme, there had been a tremendous increase in the production of shrimp. The Nellore District of Andhra Pradesh is known as the Shrimp Capital of India. There is need for Pink Revolution (Prawns) in the coastal regions of the country. Aquaculture Productivity from the FFDA supported ponds/tanks has been increased to about 2.2 tonnes per ha/per year at present. This is no doubt a good achievement; however, we cannot rest at this level of accomplishment, as much more can be done which is possible. There are advanced technologies developed to raise the production level. Programmes for augmenting freshwater production to 94 integrated farming, running water or flow through system etc. have to be taken up with varied fish species. By the year 2030, aquaculture is expected to dominate fish supplies accounting nearly over 50 percent of total fish production (FAO, 2000). About 35 percent of Indian population eats fish. Thus, annual per capita consumption of fish eating population is projected about 16.8 kg in 2010, and would rose to 18.5 kg by 2020.

Table 2.22: FFDA Achievement in Production/ha(in Kg)

Year	Area Surveyed (in ha)	Area brought under fish culture (in ha)	No. of beneficiaries	No. of farmers trained	Production /ha (in Kg)
2004-2005	1044.61	605.25	2520	1870	2050
2005-2006	853.48	623.83	1237	1494	2150
2006-2007	1182.89	554.08	3448	1602	2210
2007-2008	1174.16	578.78	3949	2519	2685
2008-2009	1355.58	656.13	2982	3037	2949
2009-2010	4213.69	2468.63	20123	8731	2960
2010-2011	5395.07	4748.22	24998	9289	2492

Source: Department of Fisheries, Government of Kerala, Thiruvananthapuram. Keralahttp://www.fisheries.kerala.gov.in/index.php?option=com_content&view=article&id=105:ffda-achievement-details &catid=42:statistics&Itemid=60

The economic issues which diversely affects the sustainability of fisheries sectors are decline in production, stagnation in consumption, overcapitalization, uneconomic operation, declining employment and productivity, poverty and barriers of trade (Ramakrishnan Korakandy,2008).

The demand for fish in future will basically be determined by increase in the number of consumers and the preference for seafood. Disposable income also plays a key role in the future. Many developed nations like Japan may face a downward revision in economic growth. An unavoidable consequence will be reduction in demand for fish. As on today Indian seafood export trade has not experienced a setback. One way to address, this problem in future is to enlarge marketing within the country, as the demand for fish and seafood is high in India.

3

MARINE PRODUCT EXPORTS AND WTO

Liberalisation, Privatisation, and Globalisations channelised massive adoption of scientific inventions and technology interventions and witnessed revolutionary challenges in all segments. And it consequently led to a rapid structural change with spectacular increase in productive development and quality of life of the people all over the world. This chapter examines the quality assurance standards followed in various countries and to studies the impact on Marine fish exports from India and Kerala. The present study also tries to assess the impact of quality assurance standards on the level and direction of exports of fish and fishery products from Kerala's seafood export processing industry.

WTO and Fish and Fishery Export Processing Industry

The World Trade Organization (WTO) was formed in 1995 to replace the General Agreement on Tariffs and Trade (GATT), which started the process of trade liberalization after the Second World War. The WTO furthers the GATT liberalization process by providing both a forum for the negotiation of new rules to liberalize international trade and an unprecedented means to resolve disputes under the existing rules (Appendix3). In November 2001, after years of discussions and concerted pressure exerted on developing countries, WTO members agreed to launch a series of multi-faceted trade negotiations known as the Doha Round (Jawara and Kwa, 2004). WTO's Doha Ministerial declaration claimed that tariffs on fish and fish products are to be significantly reduced or even eliminated. The outcomes of WTO negotiations under the Doha round, Hong Kong development round and the changing European Union regulations are likely to place new hurdles on the Seafood exports emerging from developing countries like India (Kamat and Manasvi, 2007).

The focus of negotiations is mainly on three areas, viz. Agriculture (AoA), Non-Agricultural Product Market Access (NAMA) and Services. The Agreement on Agriculture of the Uruguay Round excluded fishery products from its coverage. This means that trade in fishery products is subject to general WTO rules that apply to non-agricultural products, i.e. industrial products. NAMA represent more than 90 percent of world merchandise trade. This negotiations aim to reduce or eliminate tariffs, as well as Non-Tariff Barriers (NTB) such as import licensing systems and technical barriers to trade. The impact of trade on fish and on the environment is increasingly highlighted. These concerns will affect developing countries through negotiations on conservation, ecolabelling, and SPS measures.

The World Trade Organization is the system which provides multilateral trade opening, global trade rule-making and enforcement of these rules.WTO has 161 members as on June 2014. The WTO is a place where member governments try to sort out the trade problems they face with each other. Equal treatment is one of the basic principles of the multilateral trading system. Twenty years ago, 60 percent of the world trade was between developed countries, 30 percent was between developed and developing countries and only 10 percent were between developing countries. By 2020, trade between developing countries is expected to climb to one-third of world trade. Through supply chains, developing countries have found an accessible means to insert themselves into the global economy. Many WTO agreements contain provisions that give developing countries special rights and that allow developed countries to treat them more favourably than other WTO members. As a part of the Doha Round of negotiations, the special session of the committee on Trade and Development is reviewing these special and differential treatment provisions, with a view to making them more precise, effective and operational(WTO Report,2013).

Fisheries-related Non Tariff Barriers (NTB) in the WTO negotiations is the Agreement on Sanitary & Phyto-Sanitary Measures (SPS), and Agreement on Technical Barriers to Trade (TBT). Both agreements specifically state that members shall provide technical assistance to developing country members to ensure that the preparation and application of technical regulations, standards and conformity assessment procedures do not create unnecessary obstacles to the expansion and diversification of exports from developing country members.(Quoted from TBT agreement clause 12.6 and similar clause in the SPS agreement).The issue of TBT came to forefront during the Tokyo Round(1973-1979) of multilateral negotiations, during which time the WTO members signed the TBT agreement. The SPS agreement under the Uruguay Round Agreement on Agriculture defined SPS standards as measures taken to protect human, animal or plant life or health from risks associated with imported agricultural commodities (WTO,1995). WTO members base their national measures on the standards and recommendations of the World Health Organization's (WHO) Codex Alimentarius Commission (Josupeit, 2005).

Stringent SPS Standards have proliferated in the aftermath of the Uruguay Round Agreement on Agriculture. These standards are a major stumbling block in agriculture trade for developing countries (Gebrehiwet et al., 2007). The Agreement on the Application of SPS Measures establishes the rights and obligations of WTO members regarding measures taken to ensure food safety, protect human health from plant or animal-spread diseases, protect plant and animal health from pests and diseases, or prevent other damage from pests. Government must ensure that their SPS measures are based on scientific principles. The Agreement on Technical Barriers to Trade tries to ensure that regulations, standards, testing and certification procedures followed by WTO members do not create unnecessary obstacles to trade. Both agreements promote international harmonization in their respective areas so as to reduce the obstacles to trade. The number of regulations adopted by countries has continued to grow in response to consumer demands for safe, high-quality products, the protection of health and the need to curb pollution and environmental degradation.

The International trade in fish and fishery products in the context of the EC regulation to fight IUU fishing in third world countries is governed by the WTO Agreement on Technical Barriers to Trade. Member countries of the WTO are bound to abide by the articles of this agreement, viz. Article 2 on Preparation, Adoption and Application of Technical Regulations by Central Government bodies and Article 12 on Special and Differential Treatment of Development for Country members of the WTO Agreement on Technical Barriers to Trade that are relevant in the context of the EC regulation proposed to fight IUU fishing in third Countries. International buyers require effective application and recognized proof of enterprise system management standards such as ISO 9000 for quality management, HACCP and ISO 22000 for food safety, ISO 14000 for environmental management and SA 8000 for social accountability.

As a result, Hazard Analysis Critical Control Point (HACCP) certification has become the generally accepted requirement for the export of fisheries products. This is not only positive from a consumer perspective, as it provides a minimal standard in food safety, but it can also have beneficial conservation effects. The HACCP process deals with identifying critical contamination points in the handling and processing of food products. Because the quality of fish is very much influenced by the freshness of the product, this can have a positive conservation effect. Many countries around the world including USA, Canada, Australia and the EU countries have adopted HACCP, both in domestic and global fish trade. India has to confront serious problems by complying with the WTO agreements. It can also benefits from it by taking advantage of the changing international scenario. The main objective of the Agreement on Rules of Origin is to harmonize the rules so that all WTO members use the same criteria to determine origin in their non-preferential trade. Rules of origin are the criteria used to determine the country in which a product was made. They are used in the implementation of many trade measures, including trade statistics, the determination of customs duties, origin labeling, the application of anti-dumping measures etc.

Maarten(2013) reviewed that the WTO's achievements in the field of trade capacity building since the launch of the Doha Development Agenda (DDA) in 2001, and discussed the key challenges. The DDA has clearly created a strong footing for the delivery of the WTO's trade capacity building programmes. The evidence suggested that the beneficiaries are now much better equipped to address the challenges of the Multilateral Trading System (MTS), to take active part in the negotiations, define their strategic objectives and defend their economic and trade interests. There is no room for complacency; however, key challenges remain to be addressed. Human and institutional capacity building remain a priority for developing countries, and irrespective of the outcome of the DDA negotiations. Trade economist argues that the reduction of trade barriers allows a more efficient exchange of products among countries and encourages economic growth. A recent study by the International Trade Commission found that if the tariff cuts from the Uruguay Round were removed, the welfare loss to the United States would be about $20 billion. A study by the University of Michigan found that if all trade barriers in agriculture, services, and manufactures were reduced by 33 percent as a result of the Doha Development Agenda, there would be an increase in global welfare of $574 billion. Other studies present a more modest outcome predicting new welfare gains ranging from $84 billion by the year 2015(Fergusson, 2011).

The WTO's Ninth Ministerial Conference, held in Bali at the end of 2013, concluded with ministers approving the "Bali Package", a series of decisions covering trade facilitation, agriculture and development. The Trade Facilitation Agreement aims to streamline trade by cutting "red tape" and simplifying customs procedures. It contains special provisions for developing countries to help them implement the agreement. Benefits to the world economy are estimated to be between US$ 400 billion and US$ 1 trillion. The WTO aims to help developing countries build their trade capacity so that they can participate more effectively in the multilateral trading system (WTO Report, 2014). It is estimated that 70 percent of the benefits from successful Doha Round would accrue to developing countries as a result of increased South-South trade (World Bank, 2003). Developing countries accounts for more than half of the fishery products traded internationally. GATT, WTO, Regional Trading Agreements and the economic scenario of 1990s have brought challenges as well as opportunities for the Indian fisheries sector. Structural revolution of fish and fishery export processing sector accelerates after the liberalization of economies customized with increasing demand for value added products both in international and domestic market. Indian seafood exports increased from 1295.86 million US $ in 2010-11 to 5007.7 million US $ during 2013-14. The international trading regulations are dynamic in nature due to increasing stringent quality measures of EU, US and other developed countries. Seafood safety regulations has resulted in detention and rejections of fish and fishery products from developing countries and thereby led to widening of market destinations and increased market access to other countries.

SPS/TBT Agreements and Seafood Industry

The Agreement on the Application of Sanitary and Phytosanitary(SPS) Measures establishes the rights and obligations of WTO members regarding measures taken to ensure food safety, protect human health from plant- or animal-spread diseases, protect plant and animal health from pests and diseases, or prevent other damage from pests. Governments must ensure that their SPS measures are based on scientific principles. The Figure 3.1 shows the SPS trade concerns by subject from 1995 to 2013. A total of 368 specific trade concerns were raised between 1995 and 2013.

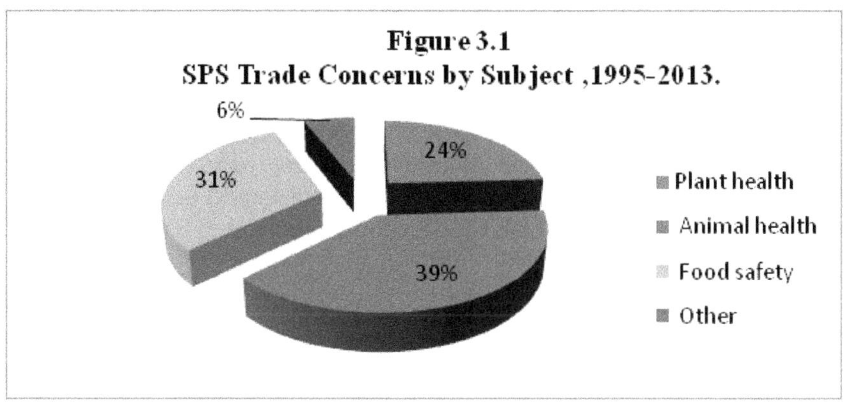

Source: World Trade Organization Annual Report 2014, Geneva.

The SPS Committee, which deals with issues surrounding food safety and animal and plant health, considered a wide range of "specific trade concerns" at each of its three meetings in 2013. A record 24 new specific trade concerns were raised and is shown in Figure 3.2. New issues covered a variety of import measures affecting trade in products from fruit and meat to seafood and swallows' nests, with actions ranging from import bans and port closures to the use of testing laboratories and outsourced certification.

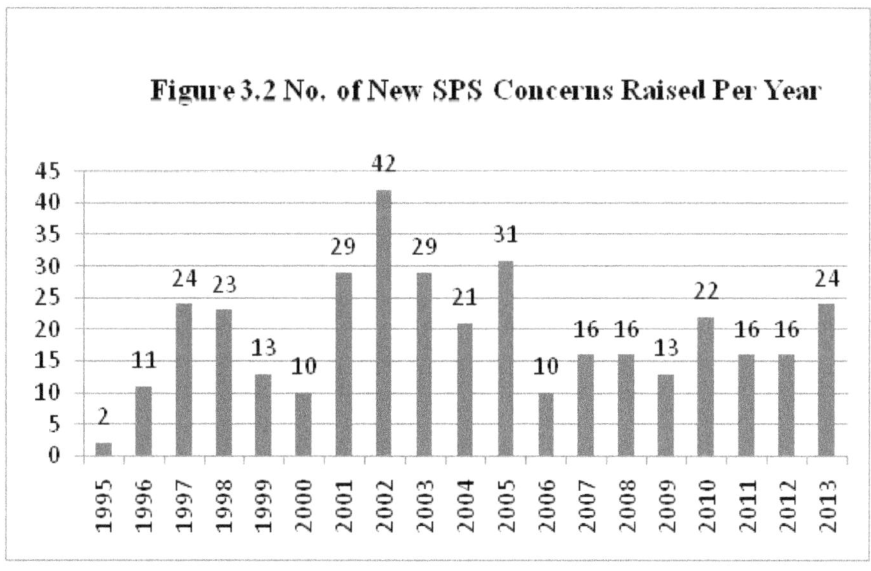

Source: World Trade Organization Annual Report 2014, Geneva.

They included US proposed food safety audits and certifications; EU temperature treatment requirements on processed meat products; China's import conditions related to phthalates, used to soften plastics, and to swallows' nests; France's ban on Bisphenol A (BPA), an industrial chemical; India's import restrictions on pork, as well as on apples, pears and citrus; EU phytosanitary requirements on pine trees from Russia, orchid plantlets from Chinese Taipei, and citrus products from South Africa; BSE-related restrictions on Brazilian meat products; and import restrictions in response to the nuclear power plant accident in Japan. The Doha Ministerial Conference of 2001 decided that reviews of the SPS should be carried out at least once every four years. A total of 10,522 people received SPS training at various levels in 268 training events.

The Technical Barriers to Trade (TBT) Agreement tries to ensure that regulations, standards, testing and certification procedures followed by WTO members do not create unnecessary obstacles to trade. The number of regulations adopted by countries has continued to grow in response to consumers' demands for safe, high-quality products, the protection of health and the need to curb pollution and environmental degradation. In an effort to curb the growing number of specific trade concerns brought before the TBT Committee, members shifted their focus in 2013 to address broader cross-cutting themes such as the use of good regulatory practices. The WTO Secretariat officially launched in October 2013 an online facility for submitting TBT notifications. The number of new or changed draft measures notified to the Committee continues to rise from 1995 to 2013 and is shown in Figure 3.3.

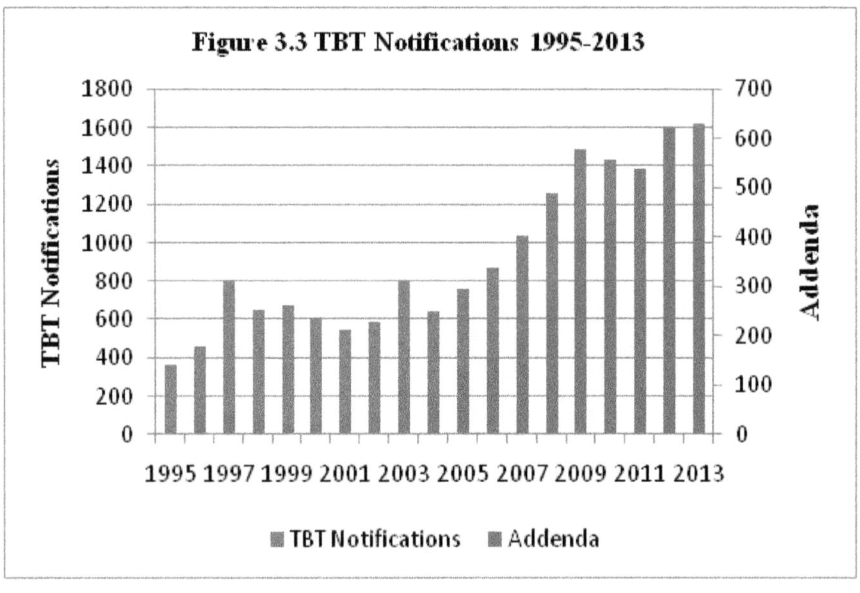

Source: World Trade Organization Annual Report 2014

Both SPS and TBT agreements have relevance in the seafood industry and have given a new direction to the international food trade. The principles of achieving harmonization of standards and equivalency in food control systems and the use of scientifically-based standards are embodied in the two binding regulations of WTO. These agreements are intended to ensure that requirements such as quality, labeling and methods of analysis applied to internationally traded goods are not misleading to the consumer or discriminate in favour of domestic producers or goods of different origin (Bostock et al., 2004). To meet the requirements of international food trade, countries have formulated quality standards to ensure food safety. These standards generally are in agreement with the conditions of international quality standards.

The SPS Agreement gives status and legal force to the standards set by the Codex Alimentarius Commission. The Codex Alimentarius or food code was created in 1963 by FAO and WHO to develop food standards and guidelines and has become a global reference point for consumers, food producers and processors, national food control agencies and for the international food trade. Generally, USFDA and EU standards are taken as models. Indian National Standards (BIS) are in more or less in tune with the USFDA/EU standards. Most of the exporting countries adopt these standards for reasons for uniformity and international acceptability. Since Indian seafood industry is mainly an export oriented industry and most of the export, particularly our frozen shrimp, cephalopods export are to EU/US/Japan markets. Adoption of the USFDA/EU standards as models have helped our seafood industry in meeting the import quality requirements of these nations very well.

In seafood, the major safety problems are microbiological, chemical and also related to the presence of antibiotic, veterinary drug residues. Spoilage and presence of certain muddy smelling compounds are the quality problems. There are strict microbiological standards, stipulating the limits of total bacterial counts, E.coli counts, and staphylococcus aureus counts and fixing zero tolerant limits for Salmonella and Vibrio Cholerae. The limits of chemical parameters are in strict compliance with international levels. Antibiotic and veterinary residues are not permitted in our processed seafood products. Our food testing laboratories are being accredited under the ISO/IEC17029 requirements. At the National level, we have already made a Food Safety Law. A National Authority will be monitoring the food processing industry for meeting and achieving the requirements under the SPS and TBT agreements.

The main purpose of TBT agreement is to encourage the use of International standards, guidelines and recommendations in technical regulation and standards, testing, packaging, marketing and labeling. Although SPS and TBT Agreements are complementary and mutually reinforcing but both differ in certain areas. SPS measure is meant to ensure food safety, environment safety and safety of plants and animals, whereas the TBT agreement is meant to ensure protection against deceptive practices. The SPS is based on scientific evidences, but the TBT is based on national security. The SPS could be area specific i.e. whether disease prone area or not, but the TBT has universal application.

Requisite for Quality Assurance Standards

From the prehistoric times, the banks of rivers and estuaries have been the nerve centers of civilization and industrialization. Increasing human activities contaminated food and impacted the toxic substances of diverse origin and nature. This led to the introduction of many hazardous non-degraded chemicals into the aquatic eco system. The alarming rate of pollutant input far exceeds that of nature's cleansing processes and has consequently resulted in an ecological imbalance (Paul, 1992). Modernisation of agriculture operation and consequent widespread and indiscriminate permeation of the ecosystem with pesticides have resulted in biological stress at all levels of organization. Pollution is the result of industrialization and technological advancement and its increase is also correlated with the rising population (Jacob, 1988). The growing demand and supply of new chemicals, including pesticides in the industrialized society in this century places increasingly higher stress on the environment. The aquatic environment is particularly sensitive to pollution because of its fragile ecology (Babu, 2001). Fishes have been identified as being very sensitive to pollutants and through them same toxicants may reach human being as well (Gopalakrishnan, 1990).

Fish is one of the most traded food commodities in the world. Seafood enjoys a virtuous status as a nutritious and healthy food. However there are growing concerns about environment contaminants in fish as well as about the poor management of fisheries in many parts of the world. The approach to ensure food safety has moved away from relying on end product inspections with accompanying laboratory analysis, towards the preventive Hazard Analysis and Critical Control Point (HACCP) approach. For years FAO has been studying the reasons for rejections and detentions of fish and fishery products at the borders of the main importing countries. The result show considerable divergence in reasons for detentions and suggest the need for harmonization of the procedures and methods that govern imports (Abdbouch et al., 2005). Food safety has been of concern to human kind since the dawn of history and the concern about food safety resulted in the evolution of a cost effective, food safety assurance method, the HACCP. PRP is considered as a support system which provides a solid foundation for HACCP. An integral management system only can assure the effective control of the quality hygiene and safety related issues (Ancy Sebastian, 2009). The main ideal behind HACCP is that it is possible to identify potential Hazard and faulty practices at an early stage in the production chain.

Adoption of appropriate technology will enhance the fisheries system many fold. Armand was the first to define a systems engineering approach to quality. Statistical Quality Control came into being in 1940s (Figenbaum, 1983). 1950s observed an era of awareness of importance of the quality control not only in the United States but also in Japan and certain European countries. Total quality control combines management methods and economic theory with organizational principles, resulting in commercial leadership. Widespread quality improvement performance in a nation's leading businesses is directly related to quality's long-term economic impact and was introduced in Japan in the 1950s, but its economy didn't flower until the 1970s (Gregory, 2005). Similarly, the United States began using quality in the early 1980s but didn't see economic success until the 1990s. The commercial success of Feigenbaum's total quality control concept is undisputable when faced with its wide number of proponents throughout the global business community. Total quality control, known today as total quality management (TQM), is one of the foundations of modern management and has been widely accepted as a viable operating philosophy in all economic sectors. Implementation of market orientation principles will lead to increase in business performance (Smitha, 2007).The business performance may be economic performance and non-economic performance that will help the seafood firms in gaining competitive advantage in exporting. A large percent of the Indian seafood firms are traditionally family owned companies rather than professionally managed firms. This would result in the promulgation of old ideas of management whereby, conflict was seen as a healthy exercise which helped to build up each department's efficiency. Quality issues matters a lot with this situation. In the current era the decentralisation of decision making facilitates the participation of the lower level of employees and builds up their motivational levels and commitment to the fishing industry.

Trade in fish products follows a rich-country-poor country dynamics because high quality fisheries products are becoming increasingly expensive and in many cases luxury food items. Essentially, high value products like crab, lobsters and shrimp go to rich country markets, while low value products flow to developing countries. Stringent international quality standards and regulations demands Kerala seafood industry to become more competitive. This study tried to capture demand side constraints related to international trade such as non-tariff barriers of quality standards and its impact on export market.

Development and implementation of Good Aquaculture Practices (GAP), Good Manufacturing Practices (GMP), Good Hygienic Practices (GHP) and Hazard Analysis Critical Control Point (HACCP) are required in the food chain step(s). Government institutions should develop policy and a regulatory environment, organize the control services, train personnel, upgrade the control facilities and laboratories and develop national surveillance programmes for relevant hazards. The support institutions (academia, trade associations, private sector, etc.) should also train personnel involved in the food chain, conduct research on quality, safety and risk assessments and provide technical support to stakeholders. Finally, consumers and consumer advocacy groups have a counterbalancing role to ensure that safety and quality are not undermined by political considerations solely when drafting legislation or implementing safety and quality policies. They also have a major role in educating and informing the consumer about the major safety and quality issues.

HACCP in the Fish and Fishery Export Processing Industry

HACCP began in the early 1960s as a result of a joint private-public venture to provide safe food for US astronauts on space missions. HACCP principles were adopted by the government of developed country in 1980's. The Codex Alimentarius Commission (CAC) has developed HACCP codes for food production, including a specific code for fisheries and aquaculture (FAO, 2011). The objective is lowering of risks rather than inspection and testing. Testing can fail to uncover contamination of some food products even if large samples are used due to the enormous variety of products and unknown probability distribution of contamination (FAO,2011). Under these circumstances, prevention of hazards is more effective.

In recent time there has been a considerable change in the methods of seafood processing. Day by day the seafood processing industry is getting more and more sophisticated and technologically advanced. Further food additives like antioxidants, preservatives, emulsifier, coloring materials and cryoprotectants are widely used. These are also problems of pesticide residues, toxic metals, mycotoxins, biotoxin residue and the like. Under all these circumstances, the responsibility of the processor has become increasingly complex and hence, there is a global shift from food quality to food safety. This has resulted in the development of a safety oriented quality system, Hazard Analysis Critical Control Point (HACCP) system. The HACCP system, which is science-based and systematic, identifies hazards and outlines measures for their control to ensure the safety of food. HACCP is a tool to assess hazards and establish control system that focus on prevention rather than relying mainly on end-product testing and inspection. Any HACCP system is capable of accommodating change, such as advances in equipment design, processing procedures or technological developments.

Prerequisite Programme

HACCP system is not a stand-alone programme. There are certain pre-requisite programmes that are very much essential to successful application and effective implementation of HACCP system. In other words, these pre-requisite programmes provide a very strong foundation for implementation of HACCP system effectively. Most of these prerequisite programmes are addressed by the Standard Sanitation Operational Procedures (SSOP) and Current Good Manufacturing Practices (CGMPs). All food Processors are required to keep a written SSOP focusing on eight areas of sanitation such as safety of water/Ice, cleanliness of contact surface, prevention of cross contamination, employee facilities, and prevention of adulteration, proper labeling, and storage of toxic substances, employee health and pest Management.

Advantages of HACCP System

HACCP is a science based system which is practicable at the industry level. It enhances food safety and reduces risk of food borne diseases. It also reduces production costs through reduced wastage. HACCP supports GMP and Quality Assurance. It helps in internal auditing which aid in inspection by the regulatory authorities. This system will augment buyer's confidence thereby promoting international trade.

Principles of HACCP System

In order to apply HACCP system in the fish and fishery export processing industry, the company should be operating according to the Codex General Principles of Food Hygiene and appropriate food safety legislation. Management

Commitment is essential for effective implementation of HACCP system. The HACCP system works on seven principles and there are twelve steps. The logic sequence for the application of HACCP system is as follows.

Assemble the HACCP Team ----> Describe the product----Identify intended used---->Construct flow diagram---->On-site verification of flow diagram ---->Hazard analysis---->Determine the Critical limit for each CCP---->Establish monitoring system for each CCP---->Establish corrective action---->Establish verification procedures---->Establish documentation and record keeping.

HACCP is based on seven principles. They are the following:-

1. Conduct hazard analysis and identify preventative measures i.e., identification of the main risks of contamination in the production and distribution process.
2. Identify Critical Control Points (CCP), i.e., areas where preventative steps can be applied.
3. Establish Critical limits at each Critical Control Point, i.e., the value of indicators that trigger corrective actions.
4. Critical Control Point monitoring requirements, i.e., the mandated procedures and their frequency for monitoring indicators at the control points.
5. Establish Corrective action to be undertaken when a critical limit deviation occurs i.e., measures to be taken in the event that the critical limits are exceeded.
6. Establish a Record Keeping System: HAACP requires records documenting the implementation of the above steps.
7. Establish Verification Procedure: Procedures for ensuring that HACCP is working correctly. Need a regular inspection and gathering of evidence on the functioning of the above steps.

Three Potential Hazards in Seafood

Physical Hazard: Glass, Wood, Stones, Metal, Insulation, Bone, Plastic, Personal effects.

Chemical Hazard: Natural occurring chemicals- Allergens, Aflatoxins, Mushroom toxins, PSP, DSP, ASP, Ciguatoxin. Added Chemicals- Pesticides, Fertilizers, Antibiotics, Growth hormones, Lead, Zinc, Cadmium, Arsenic, Food Additives, and Contaminants such as Lubricants, Detergents, Sanitizers, Paints and refrigerants. From packaging- Vinyl chloride, Printing inks, Adhesives, Lead, Tin.

Biological Hazard: Spore forming bacteria- Clostridium btulinum, Clostridium perfrigens, Bacillus cereus. Non-spore forming bacteria-Salmonella, Shigella, Staphylocuccus Aureus. Vibria parahaemolyticus, Vibrio vulnificus, Vibrio Cholerae, Entero pathogenic, Escheratia coli, Listeria monocytogenes.Virus- Hepatitis A & E.Protozoa- Rotavirus.Parasites- Taenia solium, Entamoeba histolytica.

The first detailed publication in the United States of how HACCP could be applied to the seafood industry appeared in 1977, and except for low acid canned food, few attempts were made before 1985 to apply HACCP to seafood products. At that time HACCP was recommended as the most effective way to monitor the safety of fish and shellfish. The FAO Fish Utilization and Marketing Service began in 1985 to use HACCP in its training programmes and the United States National Marine Fisheries Service (NMFS) developed a HACCP based programme for seafood (Martin et al., 1993). The Institute of Medicine, National Academy of Sciences and Committee on Evaluation of the Safety of Fishery Products published a major document on seafood safety in 1991 (Ahmed et al., 1991). The United States Food and Drug Administration issued its final rule to mandate HACCP for use in seafood processing plants in the United States effective 18 December 1997 (United States Food and Drug Administration, 1995). The European Union formally shifted to the preventative systematic approach provided by HACCP in 1991 (EEC Commission Decision, 1991). The main technical characteristic of the new inspection and quality control procedures approved at that time was the adoption and enforcement of HACCP in European Union member countries and in those countries that wish to export to the European Union. Fish is considered as a most perishable food item and has high risks of contamination so as prerequisites HACCP measures are strictly followed at all stages from catch to final consumption.

Codex Alimentarius Commission (Codex)

The Codex Alimentarius is a collection of international food safety standards that have been adopted by the Codex Alimentarius Commission (the "Codex"). The Codex is based in Rome and funded jointly by the FAO and the WHO. In the early 1960s, the Food and Agriculture Organization (FAO) of the United Nations and the World Health Organization (WHO) recognized the importance of developing international standards for the purposes of protecting public health and minimizing disruption of international food trade. Accordingly the Joint FAO/WHO Food Standards Program was established, and the Codex Alimentarius Commission was designated to administer the program.

The founders who established the Food Standards Programme and the Codex Alimentarius Commission were concerned with protecting the health of consumers and ensuring fair practices in the food trade. They felt that both of these objectives could be best met if countries harmonized their food regulations and adopted internationally agreed standards. Through harmonisation, they envisaged fewer barriers to trade and a freer movement of food products among countries, which would be to the benefit of farmers and their families and would also help to reduce hunger and poverty. They concluded that the Food Standards Programme would be a solution to some of the difficulties that were impeding free trade.

The advantages of having universally agreed food standards for the protection of consumers were recognized by international negotiators during the Uruguay Round. The SPS Agreement and TBT Agreement within the separate areas of their legal coverage, encourage the international harmonisation of food standards. Importantly, the SPS Agreement cites Codex's food safety standards, guidelines and recommendations for facilitating international trade and protecting public health.

The Codex Alimentarius is a science-based organization. Independent experts and specialists in a wide range of disciplines have contributed to its work to ensure that its standards withstand the most rigorous scientific scrutiny. The work of the Codex Alimentarius Commission, together with that of FAO and WHO in their supportive roles, has provided a focal point for food-related scientific research and investigation, and the Commission itself has become an important international medium for the exchange of scientific information about the safety of food. The standards of Codex have also proved an important reference point for the dispute settlement mechanism of the WTO.

Over the years, the Codex has developed over 200 standards covering processed, semi-processed or unprocessed foods intended for sale for the consumer or for intermediate processing; over 40 hygienic and technological codes of practice; evaluated over 1000 food additives and 54 veterinary drugs; set more than 3000 maximum levels for pesticide residues; and specified over 30 guidelines for contaminants.

International Seafood Quality Standards and Tribulations

The ocean covers 71 percent of the earth's surface and 80 percent of the value of exploited marine resources are attributed to the fishing industry. The fishing industry has provoked various international disputes as wild fish capture rose to a peak about the turn of the century, and has since started a gradual decline. Iceland, Japan and Portugal are the greatest consumers of seafood per capita in the world. Seafood industry recognized that the quality was the prime dogma in export. Proper preservation of raw material throughout its supply chain is the key to retaining its quality. Quality has to be maintained for domestic as well as export markets. Table 3.1 depicts the levels of pesticide and heavy metals allowable in fish imported into different Countries. The organizations are classified into two, compulsory legislation and voluntary standards for formulation of standards and for monitoring their quality.

Table 3.1: Levels of Pesticide and Heavy Metals Allowable in Fish Imported into different Countries

Pesticides/Heavy Metals	EU	USA	Japan
DDT	1 ppm	5ppm	3ppm
Aldrin	0.2ppm	0.3ppm	0.1ppm
Chlordane	0.02ppm	0.3ppm	0.5ppm
Fluridone	0.2ppm	0.5ppm	0.5ppm
Cadmium	0.5ppm	3ppm	3ppm
Lead	0.5ppm	1.5ppm	1ppm
Methyl Mercury	0.5ppm	1ppm	0.3ppm

Source : Deepak Shekhar, Joint Director ,EIA- Koch, Role of Export Inspection Council of India and Export Inspection Agencies in the WTO scenario, presentation made at the National Seminar on WTO & its impact on Indian Seafood Trade, organized by SOFTI and CIFT and held at Cochin on 28th June 2008

Impact of Standards on Indian Fisheries Exports

The Regression Analysis examine the effect of major trade barriers viz. international quality standards maximum residual limits and Anti-dumping duty on the marine product exports from India. It is vital to examine the impact of international quality assurance standards on marine product exports from India as they have emerged as a non-tariff trade barrier. Table No.3.2 shows the result of the Regression Analysis of international standards on Indian fisheries Export. The following Regression equations are used to find out the effect of these measures on marine product exports from India.

Equation 1 Log Tij= Log A + a1 Log CADj + Uij

Equation 2 Log Tij= Log A + a1 Log MERj + Uij

Equation 3 Log Tij= Log A + a1 Log LEADj +Uij

Equation 4 Log Tij= Log A + a1 Log CADj + a2 Mercury j+ Uij

Equation 5 Log Tij= Log A + a1 Log CADj + a2 Mercury j + a3 Leadj + a4 AntiDj+U i

Log T ij= Log of Marine Exports f*rom i to j (I =India J = import markets)*

Log A = Constant

Log CADj = Log of Cadmium Standards set by importing countries

Log Mercuryj = Log of Mercury Standards on set by importing countries

Log LEADj = Log of Lead Standards on Lead set by importing countries

AntiDj = Antidumping duty imposed by importing countries variable

Uij = Error Term

The regression results reveals that the variables such as Total Marine Export Quantity (Tij), Cadmium MRLs(CADj), Mercury MRLs (MERj), lead MRLs (LEADj), Anti-dumping duty dummy Variable (Anti Dj) are statistically significant. Total Marine Export Quantity (Tij) is having an inverse relationship with Cadmium MRLs (CADj), Mercury MRLs (MERj), lead MRLs (LEADj) and Anti-dumping duty dummy Variable (Anti Dj).

From the regression results, it is identified that the variable such as Tij and standards MRLs has an inverse relationship and it adversely affect export trends. The International framework for seafood safety and quality assurance standards is applied in international fish trade by the major importing countries are follows:

Table No. 3.2: Result of Regression Analysis of International Standards on Indian Fisheries-Export

Equation No.	Dependent (Variable)	Intercept (Constant)	Independent Variable					R^2	$AdjR^2$
			Log CADj	Log Mercuryj	Log LEADj	Anti Dj	F -Raio		
Eqn.1	Tij	3.22478 (28.29382)*	-0.46665 (-3.02987)*				9.18	0.223	0.199
Eqn.2	Tij	3.1513 (28.35089)*		-2.34494 (-3.92691)*			15.42	0.325	0.304
Eqn.3	Tij	3.277089 (27.88263)*			-0.93709 (-2.26093)**		5.11	0.137	0.110
Eqn.4	Tij	3.1513 (27.9816)*	0 126378 (0 414449)	-2.80711 (-2.2126)**			7.59	0. 329	0. 285
Eqn.5	Tij	3.1513 (28.3311)*	-0.37192 (0. .77509)	-2.80711 (-.2.24024)**	1,363841 (1.333934)		5.78	0. 33	0.30
Eqn.6	Tij	3.1513 (27.94974)*	-0.37192 (-0.76465)	-2.80711 (-2.21008)**	1.433303 (1.36754)	0.181602 (0.444589)	4.27	0.370	0.284

Source : Computed from MPEDA Government of Kerala, Kochi , World Bank Database, 1980-2014

Note :* statistically significant at 1 percent

** Statistically significant at 5 percent level

*** Statistically significant at 10 percent level

Figures in parentheses Indicates t-Statistic Value

The European Union

Geethanjali et al., (2004) pointed out that the EU has emerged as the largest regional trading block in the world with strong economic and political powers. Though EU is a strategic trading partner for India, India's share in the total imports of EU is negligible and stands at 1.3 percent. Non-Tariff barriers faced by Indian exporters in EU, particularly in the areas of standards, testing, labeling and certification requirements (Sanitary and Phyto-Sanitary measures (SPS) and Technical Barriers to Trade (TBT) are a major cause of concern. The analysis of India and EU trade shows that the overall trade balance between the two is in favour of India. It is imperative that India formulates its trade policy in such a manner that it is able to effectively combat the problem of NTB's imposed by the EU. This is possible only if India continues to strengthen its trade relations with EU and works towards gaining preferential and duty free access to the European markets as a key and strategic partner of the EU.

The EU has one of the highest food safety standards in the world. All 27 EU Member States are members of RASFF, together with the European Commission and the European Food Safety Authority (EFSA). Iceland, Liechtenstein and Norway are also full members of RASFF. In addition the European Commission and RASFF work with the WHO alert system called the International Food Safety Authorities Network (INFOSAN). All incoming information is assessed by the Commission and forwarded to all RASFF members using one of the four types of notification.

1. Alert notifications are sent when food or feed presenting a serious risk on the product available in the market and when rapid action is required.
2. Information notifications are used in the same situation, but when the other members do not have to take rapid action because the product is not on the market or the risk is not considered to be serious.
3. Border rejections concern food and feed consignments that have been tested and rejected at the external borders of the EU (and the EEA) when a health risk has been detected.
4. News Notification: Any information related to the safety of food and feed products which has not been communicated as an alert or an information notification, but which is judged valuable for the control authorities and is transmitted to the members under the heading News.

In 2013, a total of 3205 original notifications were transmitted through the RASFF, of which 596 were classified as alert, 442 as information for follow-up, 705 as information for attention and 1462 as border rejection notification. These original notifications gave rise to 5158 follow-up notifications, representing on an average about 1.6 follow-ups per original notification.

The members take action depending on the type of notification and immediately inform the Commission of the measures taken. They may, for example, withdraw or recall the product from the market. In addition, border rejections are transmitted to all border posts i.e. all 27 EU Member States, Iceland, Liechtenstein, Norway and Switzerland. This is to ensure that the rejected product does not re-enter the EU through another border post.

European Food Safety Authority (EFSA) has been formally created on January 28, 2002. EFSA covers risk assessment as well as risk communication. EFSA's role is to assess and communicate on risks associated with the food chain. It issues scientific opinions and advice to support the European Commission and EU Member States in taking effective and timely decisions on the actions that should be taken to ensure that consumers are protected. The responsibility of risk management in decision-making remains into EU's Political Institutions' hands. Two main set of legislation greatly influence seafood exports to the EU are The Common Fisheries Policy (CFP) and The Food Hygiene Legislation. The CFP establishes a legal framework for the regulation of fisheries and aquaculture activities. It has a direct impact on the EU's production capacity through fleet and quotas management. Therefore it can directly influence imports of seafood from third countries such as the US. The Food Hygiene Legislation is the EU's instrument that guarantees safe food to European consumers. It makes sure that domestically made as well as imported food complies with minimum EU's hygiene standards.

EU delegates the control of food safety to a Competent Authority in each country, who in turn ensures that exporting firms, vessels and processors are producing safe food under a system equivalent to that in the European Union. National laws are harmonised with those in place within the European Union. When the laws of any third country are harmonised, systems are monitored and at the same time control fish processing establishments and vessels are deemed equivalent, then only the exporting country is approved for export to the European Union. Individual companies are checked by the Competent Authority and, if deemed appropriate, are listed as approved in a national register, with a certification number. This register is passed to the European Commission (EC) who makes the information public via its website and other public documents and these are the so-called List I countries. Other countries that are in the process of gaining approval but are deemed to be producing safe foods are shown in List II. Shipments from List II countries are, however, subject to 100 percent border checks. India is in the List I. In addition to the certification requirements from exporting countries, the European Union operates a border inspection system to verify regularly that the European Union requirements are effectively implemented in the exporting country. European Food Safety Authority (EFSA) covers the basic concepts of equivalence and traceability.

To protect the European consumer against health threats in food or feed, in 1979, the EC introduced a European wide system for announcing and exchanging information about serious risk associated with food and feed. This system is called Rapid Alert System for Food and Feed. Heavy metals have been associated with Fishery and Aquaculture Products (FAPs) from various regions of the world for many years. The Following notifications presented in the Table 3.3 shows RASFF report 2011.Heavy metals can contaminate FAPs from the environment or from specific contaminants which might occur in water or packaging or contact material.

Table 3.3: Heavy Metal Notifications in FAPs

Element	Bivalve Molluscs	Fish and Fishery Products	Crustaceans	Cephalopods
Cadmium	2	14	4	21
Lead		1		
Mercury		76		

Source: Rapid Alert System for Food and Feed (RASFF) report 2011 http://ec.europa.eu/food/rapidalert/docs/rasff_annual_report_2011_en.pdf

Table 3.4 summarizes the number of border rejections recorded in the RASFF database for FAP originating from Ecuador, India, Morocco, Thailand and Vietnam. As we can see in the table there are few border rejections from Ecuador. The country facing the major number of border rejections is Morocco despite the great efforts deployed by the country in sanitary compliance. Conforming to HACCP standards, improvements in processing and packaging standards and development and marketing of niche products to select markets are enabled quickly in order to take advantage of gains from trade even in the short run (Dey et al., 2005) Many developing countries continue to face frequent rejections of exported seafood, with severe economic implications.

Table 3.4: Number of border rejections in five selected countries between 2009 and 2012

Countries	2009	2010	2011	2012
Ecuador	9	17	5	7
India	30	22	19	23
Morocco	34	34	41	34
Thailand	11	6	13	22
Vietnam	42	23	24	20
Total	126	102	102	106

Source: RASFF Portal http://ec.europa.eu/food/food/rapidalert/index_en.htm; RASFF databasehttp://ec.europa.eu/food/food/rapidalert/rasff_portal_database_en.htm

The high cost of compliance is a major economic obstacle to suppliers. Countries also face different economies of scale in meeting safety standards. This is exactly true for both individual processors and exporters within each country depicted with the help of Table 3.5.

Dey et al., 2005 Observed that compliance to quality specifications in India costs around 3,09,300 US dollars with a yearly operating cost of around US dollars around 41.2, which ultimately costs 0.21 to 0.28 US dollar per kg of fish.

Table 3.5: High Cost of Food Safety Compliance in Selected Asian Countries

Country	Total Investment of a Plant(000 US$)	Yearly Operating Cost of a Plant (000US$)	Cost per KG of Fish Processed (US $)
Bangladesh	277.2	34.9	0.03-0.09
India	309.3	41.2	0.21-0.28
Malaysia	795.8	113.7	
Thailand	380.9-404.8	47.6-71.4	0.01-0.014

Source: Dey et al., 2005: Food safety standards and regulatory measures: implications for selected fish exporting Asian countries. Aquaculture. Econ. Manage. 9 (1& 2): 217-236.

The EU Hygiene package has been designed to protect the health and safety of consumers as well as addressing animal welfare, plant growth and environmental protection. It follows the principles of the farm to table or food chain approach promoted by the WTO. Table No.3.6 summarizes the regulations and directives which form the hygiene package.

Table No. 3.6: The body of law and supporting acts that form Community legislation on Food Hygiene

Regulation (EC)	Title
No.852/2004	The hygiene of foodstuffs
No.852/2004	The hygiene of foods of animal origin
No.852/2004	Official controls on products of animal origin
Directive No 97/78/EC	Laying down the general principle and requirements of food law, establishing EFSA and procedures in matters of food safety
Supplemented By	
No,178/2002	The general principles of food law
No,882/2004	Reorganising official controls on foodstuffs and feeding stuffs

Source: Based upon www.europa.eu/legislation_summaries/food safety/veterinary Checks and food hygiene/184001_en.htm

The Rapid Alert System for Food and Feed (RASFF) was established originally under Article 8 of Directive 92/59/EEC (superseded by Directive 2001/95), on general product safety. Products in this case cover "any product, including in the context of providing a service, which is intended for consumers.", so the Directive is much broader than food and feed.

Table No. 3.7: Evolution of RASFF Notifications by 7 countries of origin

Country of Origin	2011	2012	2013
China	562	536	433
India	336	340	257
Sri Lanka	9	23	23
Thailand	95	119	88
Vietnam	108	74	76
Bangladesh	77	56	26
United States	113	127	101

Source: RASFF Portal http://ec.europa.eu/food/safety/rasff/docs/rasff_annual_report_2013.pdf

This Directive provides for a procedure to inform the European Union Member States when a product presents a serious risk for the health and safety of consumers. In 2013, a total of 3205 original notifications were transmitted through the RASFF, of which 596 were classified as alert, 442 as information for follow-up, 705 as information for attention and 1462 as border rejection notification. Table No. 3.7 observed that comparatively India's Notifications has reduced significantly than other countries.

United States of America

Among the principal laws associated with seafood safety in USA is the Federal Food, Drug and Cosmetic Act (the Act) of 1938, as amended (21 USC.301-392) and is concerned with practices in manufacturing processing and packing. A Federally Mandated Seafood Rule (FDA, 1995) promulgated in 1995, constitutes the basis for the sanitary procedures for processing and importing fish and fishery products into the United States of America, including Good Hygienic Practices and HACCP. HACCP programmes in the United States are monitored by the United States Food and Drug Administration and focuses on maintaining safety standards for seafood. That is, safety assurance, not quality assurance, is the cornerstone of the programme. Rejections may be due to Salmonella contamination decomposition and presence of filth. In 1994-95, USFDA estimated detention of 228 consignments of Indian Seafood. FDA's multifaceted and risk-informed seafood safety program relies on various measures of compliance with its seafood Hazard Analysis and Critical Control Points (HACCP) regulations, which describe a management system in which food safety is addressed through the analysis and control of biological, chemical, and physical hazards from raw material production, procurement and handling, to manufacturing, distribution and consumption of the finished product.

The majority of United States Federal regulatory authority and activity for seafood regulation is vested with the Food and Drug Administration (FDA) within the Department of Health and Human Service. Import alerts have been developed to communicate guidance to FDA field offices. The purpose of an import alert is to identify and disseminate import information to FDA personnel thus providing for more uniform and effective import coverage. Import alerts identify problem commodities and shippers and provide guidance for import coverage.

The Public Health Security and Bioterrorism Preparedness and Response Act, commonly known as The Bioterrorism Act of 2002 are designed to protect the U.S. from bioterrorism. The law authorizes the U.S. Department of Health and Human Services (HHS) to take action to protect the nation's food supply against the threat of intentional contamination. The Food and Drug Administration (FDA), as the food regulatory arm of HHS, is responsible for developing and implementing these food safety measures, including four major regulations.

Registration of Food Facilities: Domestic or foreign facilities that manufacture, process, pack, distribute, receive, or hold food for consumption by humans or animals in the U.S must register with the FDA no later than Dec. 12,2003. Prior Notice of Imported Food: Beginning Dec. 12, 2003, FDA must receive advance notice of each shipment of food into the U.S.

Establishment and Maintenance of Records: Persons that manufacture, process, pack, transport, distribute, receive, hold, or import food will be required to create and maintain records. FDA determines are needed to identify the immediate previous sources and the immediate subsequent recipients of food (i.e., where it came from and who received it).

Administrative Detention: Authorizes FDA to administratively detain food if the agency has credible evidence or information that the food presents a threat of serious adverse health consequences or death to humans or animals.

The Public Health Security and Bio-terrorism Preparedness and Response Act became mandatory in December 2003 and requires that domestic and foreign facilities that manufacture/process, pack or hold food for human or animal consumption in the United States of America register with FDA and submit, electronically and prior notice to FDA should be given before the shipment is due to arrive into the United States of America. It is feared by several fish exporting countries that the implementation of these requirements may disrupt fish trade flows from exporting countries into the United States of America. When exporting to the USA, the concerned importers should file the entry documents with the US Customs service within five working days prior to the date of arrival of the consignment at a port of entry. Then the US Customs service will notify the FDA about the entry of a regulated food through duplicate copies of customs entry documents and submitted copy of commercial invoice and surety to cover potential duties, taxes and penalties.

FDA conducts its seafood safety oversight activities in conformance with its statutory authorities, which have recently been expanded by the Food Safety Modernization Act (FSMA). FSMA represents the first major overhaul of FDA's food safety law in more than 70 years and will transform FDA's food safety program. FSMA closes significant and longstanding gaps in FDA's food safety authority, with new safeguards to prevent, rather than react, to food safety problems, and gives FDA important new tools to ensure that imported seafood is as safe as domestic seafood. The FDA Food Safety Modernization Act (FSMA) was signed into law in January 2011. The law, which received bipartisan support, followed a series of severe outbreaks of food borne illness. It responds to the significant burden of food borne illness in the United States each year, which the Centers for Disease Control and Prevention estimate at 48 million illnesses, 128,000 hospitalizations, and 3,000 deaths. The economic losses to industry, including farmers, are enormous, estimated at over $75 billion per year. FSMA reflects the need for a modern, global food safety system that prevents problems rather than primarily reacting to them after they have occurred. FSMA outlines broad new requirements on manufacturers, processors, packers and distributors of food, including an increase in food facility inspections and a requirement for food facilities that have a written Hazard Analysis and Food Safety Plan. The FDA is also proposing that food facilities have a written Food Defense Plan to prevent intentional contamination and tampering of food products. Registrar Corp offers services to help food companies to comply it successfully.

US FDA registration is required for all facilities that manufacture, process pack or store food, beverages or dietary supplements that may be consumed in the United States by humans or animals. Companies located Outside the United States must designate a US agent for US FDA communications.

Registrar Corp assists businesses with U.S. FDA compliance. Certificates of Registration issued by Registrar Corp provide confirmation to industry that you are fulfilling U.S. FDA registration requirements. U.S. FDA does not issue or recognize Certificates of Registration. Registrar Corp is not affiliated with the U.S. Food and Drug Administration. FDA collaborates with the President's Food Safety Working Group to modernize food safety by building collaborative partnerships with consumers, industry and regulatory partners. In addition to implementing the new FSMA authorities, FDA will continue the national residue monitoring program and recognizes the benefit of such a program to ensure that foods are not contaminated with illegal animal drug residues. FSMA directs FDA to establish a program for testing of food by accredited laboratories and will require that food be tested by accredited laboratories in some circumstances, such as in support of admission of imported food. FDA is developing the laboratory accreditation program as part of its FSMA implementation efforts.

Japan

The tragic earthquake and the accompanying tsunami in Japan is affecting World Market for food. Japan is the world's single largest importer of fish and fishery products, and in the short term, the damage to infrastructure and the disruption in transportation and electricity transmission is negatively impacting imports, distribution and consumption of chilled and frozen products. Much of Japan's fish processing capacity has already been outsourced to neighbouring countries such as China, Vietnam and Thailand, and this trend will continue. Japan's domestic production supplies 40 percent of the country's food consumption, based on calories supplied. The remaining 60 percent is imported from other countries. The safety of imported food is an important priority for the Japanese government and a strict regulatory framework is in place to safeguard the health of the nationals.

Regulatory Body

The Ministry of Health, Labour, and Welfare (MHLW) is the authority responsible for food hygiene policies and procedures in Japan. Within the Ministry, the Policy Planning Division of the Department of Food Sanitation under the Pharmaceutical and Medical Safety Bureau is responsible for the enforcement of related laws and regulations and is shown in Table No.3.8.

Food Sanitation Law

The main legislation governing the safety of imported food is the Food Sanitation Law implemented from 1948. The objective of this law is to protect the public from health hazards caused by the consumption of food or drink. Including imported food, and thereby contribute to the improvement and promotion of public health. This law relates to the content, including food, food additives, direct contact with food and food additives utensils, containers and packaging on the sale or use of hazardous substances prohibited, the benchmark specification settings, which means that the benchmark, advertising, inspection and business . In Japan's the food safety law principles and duties, the whole country follows similar regulatory principles. At present both imported food and domestically produced food products are subject to the same governing law and health and hygiene standards.

The Food Safety Basic Law

The second is the Food Safety Basic Law. The law was enacted in 2003, and aim to promote food safety and ensures that the overall strategy and is to restore the trust of the food administration.

Law concerning standardization and proper labeling of Agricultural and Forestry Products (JAS Law)

This law regulates food labeling standards to ensure that food labels are easy to understand and distinct to read. It requires food labels to be placed at a readily visible location on the packaging and specific the format of labels such as information layout dimension, Size and colour of fonts and frames. It stipulates the inclusion of the following items in food labels:

1. Name of Product.
2. Name of the raw material or ingredients.
3. Food additives.
4. Volume (net weight).
5. Best before "date.
6. Preservation method.
7. Country of origin and.
8. Manufacturer's name and address.

Table 3.8: Fishery Products and Relevant Regulations in Japan

HS Numbers	Items	Main Relevant Regulations
301	Fish(Live)	Food Sanitation Law
302	Fish(fresh or chilled except fillet)	
303	Fish(frozen, except fillet	Quarantine JAS Law
304	Fish Fillets	
305	Fish(dried, salted or smoked) or fish flour and meal	Food Sanitation Law, JAS Law
306	Crustaceans(live, fresh, frozen, dried, and salted)	Food Sanitation Law
307	Molluscs (live, fresh, chilled, frozen, dried, salted, or smoked)	Quarantine JAS Law
1603	Extract and Juice of fish, crustaceans and mollusks	
1604	Prepared fish, caviar and caviar substitutes prepared from fish egg	Food Sanitation Law, JAS Law
1605	Prepared crustaceans and Molluscs	

Source: http://www.fao.org/docrep/008/y5924e/y5924e06.htm http://www.mhlw.go.jp/english/topics/foodsafety/index.html

Seafood labeling is also governed by the JAS Law. In addition to the labeling requirements stated earlier, the JAS Law requires labeling of seafood that was previously frozen and subsequently thawed to bear the term "thawed" and cultivated products to bear the term" cultivated ", JAS law's main provisions of forestry and water quality standards and product representation obligations, was enacted in 1950 and in 2009. Table No. 3.8 shows the main quality regulations of Japan's fish and fishery products.

Import Laws and Procedures for Seafood.

Apart from MHLW, the import of seafood is also regulated by The Ministry of Agriculture, Forestry and Fisheries (MAFF) and is responsible for regulating imported food labeling and fishing operations by foreign nationals. Similarly apart from the Food Sanitation Law, the import of seafood is also subject to the provisions of Quarantine Law and Law for Regulation of Fishing Operations by Foreign Nationals. For exporting areas which are not required to be quarantined by both MHLW and the World Health Organisation, the importers must submit an application for import quarantine inspection along with an "Inspection Certification" issued by the relevant government agency of the exporting country to the Animal Quarantine Service of MAFF at the port of entry. For exporting areas which are required to be quarantined, the cargo will be subject to additional inspections. Imports from countries with known outbreaks of cholera within their territories are required by the Quarantine Law to undergo inspection for cholera bacteria. If cholera bacteria are detected, the cargo may not be imported into Japan, and must be decontaminated or disposed of. Imports, such as frozen sliced fresh fish and shucked shellfish are inspected subject to "Standards of Frozen Fresh Fish and Shellfish for Raw Consumption". According to this set of standards, the number of bacteria per gram of the inspected item must be less than 100000 and its group of colon bacilli must be dormant. Carbon Monoxide (CO) which may be applied to tinge seafood with an overlay color to give it an appearance of freshness is not a legally usable additive, if the CO is detected; the import of the cargo is not permitted.

Japan has made it mandatory that the fishery product imported to their country shall be accompanied by test reports of samples tested for banned antibiotic including the metabolites of Nitro furan issued by identified laboratories that includes EIA labs and EIC approved labs without which the consignment will be subjected to end-product testing at the destination by Health Authorities of Japan.

Much less information is available on Japanese safety and quality requirements for fish and fishery products, not least because of language constraints. Risk analysis principles are being incorporated along with spot checks at the border and with the quality control schemes that often control imports at the source. The main laws controlling entry of food products are the Food Sanitation Law and Quarantine. Other laws relevant to control the food imports are the Plant Protection Law, the Domestic Animal Infectious Diseases Control Law, Customs Law and for labeling, the Law Concerning Standardization and Proper labeling of Agricultural and Forest Products (JAS Law). When a consignment arrives at a Japanese port, a "Notice of Customs Clearance (i.e., Arrival Notification)" is sent to the addressee from a customs office, and the customs clearance procedure is initiated. The consignment is deemed to comply with the law. In these cases, the consignment can complete customs clearance, be passed onto the importer and be moved into domestic distribution.

Canada

The Canadian Food Inspection Agency (CFIA) develops and verifies compliance with appropriate product and process standards that contribute to the acceptable quality, safety and identity the fish and seafood products that are processed in federal establishments or imported into Canada. CFIA is responsible for regulating the export of fish and seafood products. To be eligible for export, fish and seafood destined for human consumption must originate from a registered fish processing establishment and meet defined standards. Canada's international trading partners often require certification from CFIA attesting that exports of fish and seafood products comply with Canadian and/or their food safety requirements. Therefore, prior to issuing a certificate, CFIA must have reasonable assurance that the product meets the importing country's requirements. Compliance is determined by reviewing the exporter's food safety and export certification systems as well as these systems' performance and/or by inspecting the product.

The type of certificate to be issued will depend on the importing country requirements (or lack thereof), and can be one of the following:

Export Certificates: These documents are negotiated between CFIA and the competent authority of the importing country. If the importing country requires an Export Certificate, no other type of Certificate can be issued unless it is also required by the competent authority.

Inspection Certificates: These Certificates are developed by Canada to attest that the product complies with Canadian Standards (e.g. Certificate of Origin and Hygiene).

CODEX certificate: developed by Canada using the model certificate provided by Codex Alimentarius, an international standard-setting agency.

If the importing country does not require a specific Export Certificate, an eligible exporter could still receive an Inspection or a CODEX certificate upon request. The exporter is responsible to confirm acceptance of these documents with the importing parties.

Russia

The main functions of Federal Agency on Veterinary and Phytosanitary Supervision are to control and supervise protection for pesticides and agrochemicals used for human protection from diseases, common for human and animals, etc. "Systems of Quality, Management of food quality on the basis of HACCP principles and general requirements" became valid from March 1, 2004 GOST R (State Standard) 51705.1-2001. Food Quality Legislation in Russia's Government Orders and Sanitary Rules are shown in the Table No. 3.9. Its necessity was dictated by the market requirements, by the modern approaches to ensure safe food production, as well as the integration of Russia into the World Trade Organization.

Table: 3.9: Food Quality Legislation in Russia: Government Orders and Sanitary Rules

Year/Order	Food Quality Legislation and Sanitation
1997	State Registration of Food Biological Active Additives
1998	Food Safety in the Russian Federation
1999	Sanitation-and-Epidemiological Human Welfare
1999	Consumers' Protection
2000	The State control and Supervision in the field of food safety and quality guarantee.
2000	The State Registration of novel foodstuffs, materials and goods
San PiN 2.3.2.1078-01	Hygienic safety and bioavailability requirements for foodstuffs, Veterinary and sanitary rules
SanPiN 2.3.2. 1290-03.	Hygienic requirements for manufacturing and realization of biological active additives to food
SanPiN 2.3.2.1940-05	Organization of children nutrition

Source: www.safefoods.nl/web/file?uuid=9f10e507-e5e7-42bf-b9d5... accessed on 31st July 2012

Revolutionary changes in Russian regulations in the field of standardization and certification of production, manufacturing and services occurred with the implementation of the Federal Law about technical regulation from July, 1st, 2003.

The activity of Federal Service for Veterinary and Phytosanitary Surveillance is guided by the Constitution of the Russian Federation, federal constitutional laws, federal laws, acts of President of the Russian Federation and Government of the Russian Federation, international agreements of the Russian Federation, normative legal acts of Ministry of Agriculture of the Russian Federation, and also by the present Act. Federal Service for Veterinary and Phytosanitary Surveillance performs its functions directly and through the regional bodies cooperating with other federal executive authorities, with the executive organs of the Russian Federation subjects, local self-government institutions, public associations and other organizations. Russia has decided on a total one-year ban on food imports, including fish, from the European Union (EU), United States and some other Western countries in response to the sanctions implemented against it over the events in Ukraine and it represented a good opportunity for Vietnam and other countries to boost seafood exports to Russia.

Australia

The Australian Seafood Standard (the Standard) reflects the seafood industries commitment to provide seafood for human consumption and that is produced in accordance with internationally recognised standards and meets the requirements of domestic and international customers and food safety authorities. The Standard define outcomes are consistent with the principles and objectives of Codex Alimentarius (responsible for harmonising international food safety standards), the Processed Food Orders administered by Australian Quarantine Inspection Service, the Food Standards Code administered by Food Standards Australia New Zealand, Commonwealth State and Territory food safety regulations. The seafood business takes steps to ensure that seafood is not, or does not become, contaminated by biological agents, chemicals or substances foreign to seafood and takes reasonable measures to ensure that seafood handlers handle seafood in a safe way.

AQIS, the Australian Quarantine and Inspection Service, is part of the Australian Government Department of Agriculture, Fisheries and Forestry (DAFF). AQIS sets the guidelines for certification that require Australian seafood exports to meet strict food safety and hygiene standards. The Food Standards Code includes specific provisions to ensure seafood safety for the domestic market, including imported seafood. The regulatory framework provides for consistent and risk-based management of seafood safety. The Fish Exports Program provides operational, policy and technical advice to the export seafood industry to assist in maintaining international market access. The provision of trained inspection staff supports this objective. The Fish Exports Program manages and facilitates exports according to legislation, most important of which are the Export Control Act 1982, the Export Control (Prescribed Goods - General) Order 2005 and the Export Control (Fish and Fish Products) Orders 2005.

In order to export, land-based establishments and vessels that undertake processing, as defined in the Export Control (Fish and Fish Products) Orders 2005 are required to be registered with AQIS as per the Export Control (Prescribed Goods-General) Order 2005. It is a condition of registration that the establishment has an AQIS approved quality assurance system in place, called an Approved Arrangement (AA), which meets the requirements of the Export Control (Fish and Fish Products) Orders 2005. The AA replaces the Food Processing Accreditation (FPA) and Approved Quality Assurance (AQA) systems of inspection provided under old legislation.

As per the revised quarantine measures adopted by Bio- security Australia under Policy Memorandum 2007/6 all the uncooked peeled prawns shall be accompanied by a health certificate issued by the Competent Authority as per the format specified by them. These consignments will be quarantined at the destination by the Australian Quarantine and Inspection Service (AQIS) and each batch of uncooked peeled prawns will be tested for White Spot Syndrome Virus (WSSV), Yellow head Virus (YHS), Infectious Hypodermal and Haematopoietic Necrosis Virus (IHHNV). Consignments found positive for virus will be re-exported or destroyed.

China

In order to enhance seafood quality control, China has currently carried out the principle of security management. The national authorities have issued a serial law, regulations and standards related to seafood quality control such as The Product Quality Law, The Food Hygiene Law, The Standardization Law and The Commodity Inspection Law, and instructed The Ministry of Agriculture, The State Technique Monitoring Bureau and The State Import-Export Commodity Inspection Bureau work together for seafood quality control. For those exported seafood, China primarily adapts the existing international standards or the standards and regulations by targeted countries in order to meet their demands. For example, the processing method for baked eel exported to Japan is based on the requirements by Japanese Government and the traders. To ensure the quality of exported seafood, the processing enterprises have been gradually instructed to follow the international quality management system, such as HACCP created by Food and Drug Administration (FDA) and the National Oceanic and Atmospheric Administration (NOAA), the Seafood Quality Management Criteria by Canadian Fishery and Marine Ministry and the related regulations by the European Union.

India has signed an agreement with China's General Administration of Quality Supervision, Inspection and Quarantine (AQSIQ). AQSIQ is a ministerial administrative organ directly under the State Council of the People's Republic of China in charge of national quality, metrology, entry-exit commodity inspection, entry-exit health quarantine, entry-exit animal and plant quarantine, import-export food safety, certification and accreditation, standardization, as well as administrative law-enforcement. Among potential markets, the agency targets China, expecting an increase in the trade between the two countries. Peoples Republic of China has issued new hygiene and quarantine regulations for the import of fishery products to China. A new bilingual health certificate has also been specified with effect from May 2006.

Chinese Food Safety Law of 28 February 2009 and of the bills introduced in the United States Congress in 2009, namely the Food Safety Rapid Response Act and the FDA Food Safety Modernization Act, which may undoubtedly contribute to safeguarding the health of consumers of large sections of the world population (especially in consideration of the fact that these initiatives concern two of the most populated countries in the world), and hence protect global health.

International Seafood Quality Standards in Other Countries.

HACCP programmes in other countries often include quality standards as well as safety standards in programme design. New Zealand Food Safety Authority has issued a revised standard for the import of bivalve mollusks. New Zealand Food safety Authority has made it mandatory that the production and relaying areas of bivalve molluscs shall be approved and monitored by the Competent Authority.

Hygiene Codes, BRC (British Retail Consortium), SQF (Safe Quality Food) and EHEDG (European Hygienic Equipment Design Group) are the other Food Safety Systems in the international market. **Hygiene Codes** cover the elements which are laid down in the legislation to comply with basic matters on Hygiene and Good Manufacturing Practices and provide the conditions to ensure the safety of food products. Hygiene Codes assist and inspects the relevant items of the implemented system. **BRC** originated in the United Kingdom. The BRC has developed the Technical Standard for those companies supplying retailer-branded food products. The standard has been developed to assist retailers in their fulfillment of legal obligations and protection of the consumer, by providing a common basis for the inspection of companies supplying retailer- branded food products. SQF 2000 is a HACCP quality code designed in Australia. **SQF** focuses on food safety and quality issues including GMP, SOPs (Standard Operating Procedures) and HACCP and its compatible with the ISO (International Standard Organisation) 22000 standard. The **EHEDG** is a consortium of equipment manufacturers, food industries, research institutes and public health authorities, founded in 1989 with the aim to promote hygiene during the processing and packing of food products. European legislation requires that handling, preparation, processing, packaging, etc., of food is done hygienically, with hygienic machinery in hygienic premises (EC directives 98/37/EC and 93/43/EEC).

It would be advantageous to harmonize not only the systems of border control, but also the standards, criteria and testing methods. This need has been recognized internationally, and the work of the Codex Alimentarius Commission is very important in this area. The export to the European Union still poses serious threats due to the quality aspects raised by the importers and the characteristics of a buyer market. The upgraded cost of products is a major concern in fish industry worldwide. Seafood Industry is facing continuous changes in the specific food safety requirements and the associated conformity assessment procedures applied to fish and fishery product imports in the main markets served by India. Quality related problems have become dominant in the seafood export processing industry in Kerala. Seafood export processing industry in Kerala has to operate with higher cost to ensure compliance with new requirements in the global market. This has resulted in the rejection of structures and forced market diversification to comparatively less remunerative markets observed in the post WTO phase and demonstrates a decline in the producers' surplus (Parvathy, 2012). The main constraints in adoption of quality assurance systems are lack of awareness, high cost for adoption of quality assurance systems, no legal obligation to adopt the quality assurance system and lack of time frame for implementation of quality assurance systems.

Historical View of Seafood Quality Assurance System in India

History of Quality control and Inspection system in India are In Process Quality Control (IPQC), Quality Control Inspection in Approved Units (QCIA) and Modified in Process Quality Control (MIPQC) (Shassi, 1998). Product certification system shifted to quality certification system in the international market in the early 1960s.Quality Assurance was introduced systematically with the enactment of the Export (Quality Control and Inspection) Act 1963. In the year 1964, Export Inspection Council was established. Development of Quality Assurance System in India for export of fish and fishery products was introduced in the year 1965 when frozen and canned shrimps were brought under the purview of the Export Act 1963. Fish and fishery products also come under the purview of quality control and pre-shipment inspection scheme. The testing for bacteriological factors was introduced in the year 1973. In 1978 in process quality control system was introduced in the beginning prescribing the minimal requirement for raw material manufacturing process, end product testing and preservation and packaging of the final product.

In the beginning of 1995 the HACCP system started giving way to new quality concepts based on assessment of risk to human life and health particularly risk arises on account of harmful chemicals, antibiotics, pesticides, residues heavy metals and working places. The growing demand and supply of new chemicals including pesticides in the industrial society in this century places increasingly higher stress on the environment. The aquatic environment is particularly sensitive to pollution because of its fragile ecology (Babu V, 2001). Many consignment of squids exported from India were rejected/detained by some members of European Union on the ground that the samples had higher levels of cadmium or salmonella contamination (Prafulla, 2002). He distinguished five different source from which metal pollution of the environment originates; weathering of rocks, industrial processing of ores and metals, effluents from metallurgical industries, leaching of metals from garbage solid waste dump, sewages, dumping of waste from tankers and cargo ships ,other anthropogenic sources and also exists bioaccumulation of pollutants through food chain. With the introduction of the EU directives on seafood's, EU/91/EEC and US regulation 1997, monitoring of chemical hazards, toxic metals, pesticides etc becomes mandatory.

Export of frozen cephalopods commences in 1973 and there was a phenomenal growth thereafter. Initially Burma, Singapore, Japan Italy and UK were the importers and now more than 50 countries import various products from India. Cephalopod resource is regarded as one of the most promising seafood resources because of high sales value supported by strong demand. Dry squid is highly priced than fresh or frozen squid (Mohan, 2007).

Seafood quality is affected by handling, gutting, processing and storage temperature. For the seafood industry, the monitoring of indicator compounds such as biogenic amines, associated with seafood safety is as important as quality assurance (Beena, 2008).Modern fish processing technologies such as modified atmospheric packaging, refrigerated storage, reduction in time between catch and consumption along with the use of natural antioxidants can promote the quality fatty fishes for human consumption. Biogenic amines have been studied extensively because of their involvement in food born diseases. Among the biogenic amines histamine is potentially hazardous and is a causative agent of histamine intoxication associated with the consumption of seafood (Morrow et al., 1991). It is seen that many countries importing frozen products from India are reprocessing them and marketing in their country and also re-exporting as consumer packs adding value to the products.

Quality is one of the aspects which are constantly changing with the improvement of technology and innovations. As and when importing countries introduce new regulation, seafood exporter must be cautious to new regulations and comply with them in addition to ensuring the implementation for the betterment of the export. All the seafood need to be handled with utmost care from the capture to the site of shipment. Any lapse in handling processing storage and distribution can lead to deterioration of quality. It has been estimated that good quantity of landings are wasted due to improper handling on board fishing vessels and at the landing centers. In order to minimize this wastage and to meet the quality of international standards and thereby increase the foreign exchange earnings, several measures have been taken by MPEDA. Arranging delegations from India to negotiate quality issues, inviting and arranging the visit of foreign health officials, training QC personal abroad and in India, assisting industry to implementing HACCP, assistance Schemes to strengthen quality control, Subsidy assistance for setting up mini laboratories, assistance to seafood processors to establish captive impendent pre-processing center, improvement of fish landing centers, setting up of labs equipped with LS MS-MS and Liquid Chromatography with Tandem Mass Spectrometer.

Bureau of Indian Standards (BIS)

Bureau of Indian Standards is the National Standards Body of India that covers product quality certification, consumer affairs and development of technical standards. MSMEs can strengthen their product quality and thus improve their business competitiveness by implementing these standards. Further, Ministry of Commerce Under the "National Programme for Organic Production" has also prescribed certain National Standards for Organic Production. The activities of BIS in the field of agro-processing are twofold. First one is formulation of Indian standards and the second one is their implementation through its voluntary and third party certification system. BIS has on its record over 700 Indian Standards related to food-grains and their products, bakery and confectionery items, sugar, edible starches and their products, processed fruit and vegetable products, protein rich foods, stimulant foods like tea, coffee and cocoa, alcoholic beverages, spices and condiments and food products of animal origin like milk and meat, fish, poultry etc. These standards, in general, cover raw materials permitted and their quality parameters, hygienic conditions under which the products are manufactured and packaging and labeling requirements. The standards also prescribe, where required, freedom from toxic substances and contaminants. In addition to the product specifications, which include both raw materials and final product, the Bureau of Indian Standards has brought out standards on glossary of terms for various industries and hygienic codes applicable to most of the food processing industries. To ensure that processed foods are free from pathogenic or spoilage micro-organism, limits are gradually being introduced in various specifications. In addition, separate standard methods of test for the sensory parameters have been laid down. Containers are as important as the contents, as these may impart toxic elements to the food products and could pose potential dangers to health and safety, if they are not of the requisite quality. Indian Standards covering thermoplastics namely, polypropylene isomers and ethylene acrylic acid have been published which describes the requirement of the particular thermoplastic and necessary additive along with their limits. Informative labeling is also a very important area and the level should contain sufficient information to enable the consumers to know about positive nutritional characteristics such as protein, fat, dietary fiber etc., negative characteristics such as pesticides, residues, toxins etc. as also information regarding ingredients used, food activities, net contents etc. In this area, the Bureau has brought out a Code of Practice for labeling of Pre-packed foods covering general guidelines for labeling and guidelines on claims and nutritional labeling. However, in this area, much work still remains to be done.

National CODEX Contact Point (NCCP)

The Food and Standards Authority of India, (Ministry of Health and Family Welfare) has been designated as the nodal point for liaison with the Codex Alimentarius Commission. The National Codex Contact Point (NCCP) has been constituted by the Food Safety and Standards Authority of India for keeping liaison with the CAC and to coordinate Codex activities in India.

Responsibilities of NCCP-INDIA

The NCCP shall have the certain responsibilities to discharge its core functions. NCCP would like to undertake secretariat responsibilities to the National Codex Committee. It acts as the contact point for the country for maintaining liaison with the Codex Secretariat in elaborating international food standards. Their main responsibility is to collect, procure and analyze data for elaborating international food standards. They will always keep track of international food standards work and give comments and data to ensure that international food standards are practicable for local manufactures and do not hinder exports of food. They would undertake study and research work to solve any problem resulting from the elaboration of international food standards. They are always encourage food manufacturers to improve quality and hygiene management to meet requirements of international food standards and disseminate information of food standards and food laws to relevant government agencies, primary producers, manufacturers, exporters, consumers and concerned organizations.

NABL accredited Aquaculture Pathology Lab.

The Central Aquaculture Pathology Laboratory of Rajiv Gandhi Centre for Aquaculture (RGCA), Marine Products Export Development Authority (MPEDA), became the first aquaculture pathology laboratory in the country to be accredited by NABL (National Accreditation Board For Testing & Calibration Laboratories) on May 30, 2014. Rajiv Gandhi Centre for Aquaculture, the Research & Development arm of MPEDA established a state-of-the-art Central Aquaculture Pathology Laboratory at its headquarters in Sirkali, Tamil Nadu in September 2011. The laboratory has been serving aquaculture industry of the country by providing timely and reliable diagnosis of various diseases encountered during the culture of finfish, shrimps and freshwater prawns.

The laboratory is equipped with all the latest disease diagnostic tools and has three component units for Molecular Pathology, Histopathology and Microbiology which together helps in diagnosis of various diseases encountered by the aquaculture industry. The Molecular Pathology Laboratory has been routinely engaged in screening/diagnosis of 16 Crustacean (shrimp/crab) pathogens including World Organization for Animal Health (OIE) listed pathogens besides 3 finfish pathogens. In addition, the laboratory conducts periodical and need-based aquaculture disease surveillance in India for the benefit of the seafood export industry. All the known diseases affecting the shrimp/prawn/ fish are diagnosed at this laboratory which includes the recently emerged dreaded shrimp disease viz. Early Mortality Syndrome (EMS)/ Acute Hepatopancreatic Necrosis Disease (AHPND).The laboratory has been successfully participating in Proficiency testing conducted by Aquaculture Pathology Laboratory, University of Arizona, USA (an OIE reference laboratory for crustacean diseases). Due to high perishable nature of fish various preservation methods like drying, curing, canning, freezing etc have evolved over a period of time. Although each methods are aimed at preventing changing the nutritional value and sensory quality of food by controlling growth of micro organisms and obviating contamination by adopting economic methods (Sreenath ,2009).

High Pressure Processing Foods (HPP)

HPP is a method of food processing where food is subjected to elevated pressures (up to 87,000 pounds per square inch or approximately 6,000 atmospheres) with or without the addition of heat, to achieve microbial inactivation or to alter the food attributes in order to achieve consumer –desired qualities. Pressure inactivates most vegetative bacteria at pressures above 60,000 pounds per square inch. HPP retains food quality, maintains natural freshness, and extends microbiological shelf life. The process is also known as high hydrostatic processing (UHP). High pressure processing causes minimal changes in the fresh characteristics of foods by eliminating thermal degradation. Compared to thermal processing, HPP results in foods with fresher taste, and better appearance, texture and nutrition. High pressure processing can be conducted at ambient or refrigerated temperatures, thereby eliminating thermally induced cooked off-flavors. The technology is especially beneficial for heat sensitive products.

Fish exporters from India have faced ongoing challenges in meeting food safety requirements imposed by country importers. This study focuses on fish and fishery products in the state of Kerala. There are significant differences in the food safety requirements among the main Indian fish importers, namely the EU, the US, and Japan. Border inspection of hygiene is the dominant food safety control in Japan. For both the EU and the US, fish imports must be processed in facilities with standards that match their domestic facilities. Additionally, the EU requires that the exporting country government has regulations in place that comply with EU standards. To date, Kerala has invested $13.5 million to comply with EU hygiene standards, but only limited improvements have been made so far. Stricter standards have had the greatest impact on the preprocessing sector, forcing many preprocessing plants to shut down. On the positive side, however, Kerala has raised hygiene standards at these facilities. Evidence shows that the dominant trend in the next few years will be the consolidation of numerous minor processing companies into several major ones that will dominate the sector. Kerala's main challenge for the future will be to apply stricter food safety standards while maintaining a competitive advantage in the world market.

Even though the present expansion of fisheries sector in the State is much better than the past, there is much scope for modernization and diversification of the existing scenario. There is great need to introduce deep sea fishing technologies and diversification of existing fishing fleet, addition of the recent trends in fish processing industry to cope with the international standards, efforts for boosting the coastal & inland aqua culture sector and introduction of cold storages and cold chain and modernization of the fish markets. Many Consignments of squids exported from India were rejected/detained by some members of European Union on the ground that the samples had higher levels of cadmium or salmonella contamination (PrafullaV, 2002).

Food Safety and Quality Problems in Fish Sector

The major food safety problems encountered for fish and fish products are microbiological contaminants due to lack of hygiene in the production process and residues of antibiotics, metal contaminants, parasites and problems due to temperature changes caused by a broken cold chain. Prohibited use of antibiotics has caused major problems in recent years. Antibiotics are available for the treatment of diseases in shrimp and often are used in fish farms in China, India, Thailand, and Vietnam. However chloramphenicol and nitrofuran are prohibited in Europe and consequently, a zero level tolerance has been enforced.

Quality Revolution in Kerala's Seafood Industry

After the Sanitary and Phyto sanitary International regulations came into force, stringent quality controls for seafood products, were made as a result and fish and fishery products are subjected to a number of food safety requirements related to hygiene and stipulated microbiological and chemical containments, because these products are treated as high-risk in food major markets that have undertook, over whelming changes in the safety requirements, due to factors such as advances in scientific knowledge, consumer concerns and political pressures. The demand for stringent and high hygienic standards in the production and processing facilities have greatly increased, after the stipulation of Hazard Analysis Critical control point (HACCP) in 1993 by the United States Food and Drug Authority (UDFDA), Codex Alimentarius Commission (CAC, 1996), such as ISO 9000 and other European Community directives and the EC ban on Indian marine products in 1997(Rajasenan, 2005). Implementation of quality units control measures like HACCP is highly capital intensive, which cannot be borne by small seafood processing. The cost of implementation of HACCP should be realized from the market price of products, which usually happens in the case of export oriented units (Hussan et al., 2012).

EU legislation requires strict, restrictions on landing of fish, structure of auction and wholesale markets and processing facilities, like construction of walls and floors, lighting, refrigeration, ventilation, staff hygiene etc, processing operations, transportation, storage, packaging, checks on finished products including visual, chemical and microbiological parameters, laboratories and water quality. There is a need for greater quality consciousness in the country. A widespread educational and awareness generation programme on quality standards should be a necessary condition for solving quality issues. Unless we achieve both productivity and quality revolutions, our farm products will not be globally competitive.

Quality Standard Issues in Kerala's Seafood Industry

Kerala's economic future and prosperity depends heavily on agriculture, including plantation crops, animal husbandry and fisheries. Kerala's shrimp exports have largely landed in USA and Japan. EU is the primary market for cuttlefish and squid exports. Therefore we should always be updated with the changes coming in the International Quality Standards, particularly with reference to market access and product diversification. Apart from this we should also give high focus on improvements in productivity, quality, value-addition, cost reduction, post harvest technology etc. There should be a centralized agency to monitor hygiene and health practices for both domestic and international market. In order to improve the productivity, profitability and stability of Kerala's seafood industry, there is need for the following revolutions, viz. Productivity revolution, Quality revolution, Value addition Revolution (ready to eat and ready to cook products etc) for both home and global trade. Seafood Export Processing units require financial and technical assistance to implement International Quality Standards.

India gains global leadership by getting a crucial poor-rich country imbalance corrected on a multilateral forum. The trade facilitation agreement will improve competitiveness of Indian industry both in domestic trade and international trade. India recognizes the benefits of a trade facilitation agreement, which aims at simplifying customs procedures and reducing transaction costs. Kerala's seafood industry will also boost up due to Bali package which facilitate seafood export to compete with international standards.

Constraints of Seafood Industry in Kerala

Seafood processing and marketing is becoming increasingly competitive globally. The emerging global thrust to adoption of hygienic production practices for adhering to the World Trade Organization standards; compliance to Hazard Analysis and Clinical Control Point (HACCP) by seafood exporters is a prerequisite. The strict enforcement of quality standards by the US and the EU led to hasty implementation of Codex standards and HACCP regulations, which entailed additional cost by way of both fixed and variable costs, which ranged between 10 to 20 percent of the total cost [Ouseph, 2005].

Fish is a highly perishable product. Fish is sold and priced on freshness criteria. The fishes are brought on shore at designated landing sites where it is graded by fish inspector into different price groups based on the freshness using sensory analysis (Chebet, 2010). Many exporters in Kerala have experienced border detentions in the EU. Indeed, the risk of rejection has become a normal part of doing business in these markets. Thus, there is a standard clause in the Letter of Credit (LC) used by EU buyers stating that payment is "subject to health inspection at the EU port of entry." Overall, therefore, the EU is regarded as the highest risk market. (Rajasenan, 2005). The high level of rejection of fish and fishery products in international trade due to 'filth' indicates that there is room for improvement in production and distribution systems (Valdimarsson, 2011).

Strategies of Seafood Export Processing Industries

In order to meet the international standards, various governmental agencies like Marine Products Export Development Authority (MPEDA), Export Inspection Council of India (EIC), Central Marine Fisheries Research Institute (CMFRI) and Central Institute of Fisheries Technology (CIFT) are particularly engaged in the improvement of the seafood industry in India by providing training, subsidies and financial assistance. Technical constraints can be solved by capacity building with the assistance of international bodies. The demand for fish and fishery products are increasing considerably in the country, both in domestic and exports fronts. Kerala should improve quality and safety standards by adaptation of EU and other international legislation to local conditions and requirements. Economic Incentive to be provided by the Government and banks in the form of subsidies, tax deductions, loans and organize award schemes to recognize and reward organizations excelling in HACCP implementation. Low level of awareness of hygiene among suppliers, traders, distributors and retailers can be eliminated by creating awareness thorough seminars and training at national and state levels.

Quality Control

The hygiene and sanitation conditions in most of the harbours and fish landing centers are below the normal specifications. This is partly due to inadequacies in the design and construction of the facilities and partly due to poor maintenance. The user groups are largely responsible for the poor state of hygiene and sanitation. Accepted standards of hygiene and handling of fish demand that these facilities be maintained properly and contamination of fish be kept down to a minimum. The fishing harbours need to be modernised to meet minimum international standards for fish quality assurance. Special design approaches should be adopted to meet the requirements of standards laid down by Hazard Analysis and Critical Control Points (HACCP) and ISO 9000. If these requirements are not met in the immediate future, the marine products exports may face trade restrictions, since most of the importing countries have stringent hygiene and sanitary conditions. Further, a greater awareness is necessary for the fisher folk on the importance of hygienic handling and preservation of fish as also personnel hygiene in improving the quality of landed fish and prevention of loss and wastages.

One of the major challenges facing exporters of fish and fishery products in developing countries is the stricter food safety requirements in major industrialized countries. The cost of compliance with these requirement can be high (Cato 1998, Cato and Lima, 1998), in some cases prohibitively high. The resultant impact on the structure and modus operandi of supply chains can have significant economic and social consequences for developing countries. In many cases, this impact reflects the fact that investment in upgrading supply chains and regulatory systems has not been correlated with the expansion of exports.

SEAI has identified the areas for development and implementation of Research and Development of New Projects. There is need for training in new technology and inviting overseas technical experts to India. Government should provide overseas training to technologists of Indian Seafood Industry in quality control section and also monitor the Seafood Quality in landing and pre-processing centers. Upgradation of seafood quality should be carried out by providing infrastructural facilities like pre-processing centers and setting up of mini labs for quality assurance. All the seafood export industries should meet the standards for compliance for export of fish and fishery products to various developed countries based on standards, norms and regulations prescribed by such countries. In recent years food safety has become a subject of increasing importance internationally. And the recent trends include an increased emphasis of food safety regulations in international trade, a tightening of standards by major importing countries, a reorientation of private sector quality assurance techniques towards preventive management and a corresponding shift in the role of EIC toward process-based standards including mandatory HACCP.

Quality Problems in Indian Seafood Export Processing Industry

During the eighties and the nineties, Indian Exports of marine products were under severe strain owing to the quality problems faced in the importing markets and the detention by the USFDA. Indian products were generally perceived to be of poor quality. During this period the European Union introduced stringent quality measures on imported fishery products vide its directive 91/493/EC dated 22nd July, 1991.Consequently end product testing was replaced by in-process quality control a system which required extensive changes in the processing plants. As per the new guidelines of the Export Inspection Council of India had also notified the requirement stipulated by the European Union vide notification SO 730€ dated 21/8/1995. Conformity to these specifications necessitated heavy investment in modification of seafood plants. To make matters worse the European Union imposed a total ban on Indian seafood during the year 1997 which came as a severe blow to the export industry. However, this triggered off a stupendous effort on the part of the industry with the assistance of MPEDA to upgrade seafood plants to the EU standards.

For the renovation and modernization of plant and machinery as per EU regulations, the units in India had to spend Rs 150 lakhs to Rs. 200 lakhs per plant. Due to lack of financial support from Banks and Financial Institutions and the time taken for completion of the various formalities at the Government level and at European Union, many of the unit's got the approval 2 to 3 years after the European Union ban. During this period the companies were unable to do export business adequately. This was a common factor in almost all units in India. Many of the units are not renovated due to lack of support from Financial Institution and banks. Based on the advice of the Union Government, MPEDA gave a report on the implementation of the Rehabilitation package for the seafood industry, to the Ministry of Commerce and Industry on 27.02.2006 based on the extensive study conducted by it and a recommendation for settlement of dues by the banks and for revival of the seafood industry.

Due to the reasons above which are beyond their control nearly 150 small and medium sick sea food exporting companies all over the country are under the threat of liquidation by the banks and financial institutions. They have approached appropriate authorities like MPEDA, IBA, ECGC and the Chief Ministers of Kerala, Andhra Pradesh, Gujarat and Maharashtra. They have also appraised their grievances to the commerce and finance ministers. Even though the marine industry comes under the Commerce Ministry, they have to deal with different ministries like Ministry of Finance, Ministry of Agriculture, and Ministry of Food Processing for the intricacies and interrelated problems.

Survey Data Analysis of International Quality Standards and Compliances

Out of the 55 surveyed seafood firms, 27 firms have only 1 approved technologist in their Quality Control Laboratories, 18 firms have 2 approved technologist, 4 firms have 3 approved technologist, 4 firms have 4 approved technologist, 1 firm has 6 approved technologist, and 1 firm has 10 approved technologist in their Quality Control Laboratories. Seafood industry's export performance depends on the quality of fish and fishery products exported in the international market and the ways and methods chosen to meet the international standards. For meeting the quality requirement of seafood industry efforts and contribution of approved technologist in QC Laboratories services are more relevant. Safety Standards adopted by seafood export processing plants are USFDA (Codex), EU, BRC, ISO2200, HACCP and FSSAI shown in Table No. 3.10. Among this Safety Standards HACCP is mandatory. 48 survey units have European Union approval number and can be export to EU Countries. Food Safety and Standard Authority of India (FSSAI) was formed in August 2011. Out of the 55 surveyed units 44 plants have initiated the licensing procedures and have been mandated by the Food Safety Standard Act, 2006. These various standards can be viewed as non-tariff barriers to trade and becoming more restrictive.

Table No 3.10: Seafood Safety Standards

Standards	No. of Units
USFDA	30
EU	48
BRC	24
ISO 22000	34
HACCP	55
FSSAI	44

Source: Survey Data

Sources of information on changing standards and regulations in importing country can be attained from seafood associations, government agency, foreign affiliates and foreign buyers. Survey indicated that SEAI and Government agencies have good significant role to provide updated changing standards information to the concerned seafood export processing firms depicted in table 3.11.

Table No. 3.11: Sources of Information on Standard Sources of Information

	No. of Units
Seafood Export Authority of India(SEAI)	46
Government Agency	49
Foreign Affiliates	16
Foreign Buyer	43

Source: Survey Data

An analysis of the survey of the firms revealed that out of the 55 units 45 seafood companies purchased raw material for export only i.e. their extend of export is 100 percent. 3 surveyed units have 95 percent of the raw materials meant for export, 3 surveyed units have 90 percent, 3 surveyed units have 80 percent and one surveyed units have 60 percent of these products. This is due to the change in the consumption pattern in the domestic market.

Of the 55 surveyed units, 31 units reveled that they are confronted with import detention and rejection for the past five years. 18 units pointed out that they adhere to the international quality standards. Survey indicated that 65 percent of the rejection and detention are from the European Union. The main reasons for detention and rejection are due to the presence of biological, chemical and physical content viz. salmonella, antibiotic, cadmium and heavy metals. Other issues like provision of inadequate information, coding mistakes, default in packaging and deficiency in labeling, documentation delay and lack of letter of credit and adultery. Loss of demurrage cost varies between 3 lakhs to 3 crores. The loss or demurrage can be overcome, if it can be resolved by repacking or relabeling and then diverting the consignment to a next destination if possible through agents available in the landing locations. But if the rejection is due to antibiotic issues, consignments will not return to the original destination and such consignments are destroyed within the importing destination border which incurs huge losses to the exporter and makes them almost bankrupt. This is one of the most serious problems faced by the exporters. Another important issue noted is that international standards vary from one market to another quality standard due to varying modern scientific techniques and importing requirements. Seafood export industry had to promote harmonization of quality assurance standards among the trading nations to avoid paradigm shift.

Expenses for complying with Quality Standards

Out of the 55 surveyed units, 46 units availed subsidies provided by MPEDA for complying with quality standard improvements. Upgrading the international quality assurance standards is an indispensable part of seafood export processing industry. The European standards are more stringent than other standards. The Seafood Exporters Association of India claim to have spent US$25 million on upgrading their facilities to meet the regulations. HACCP training is mandatory for all the seafood export processing industry. Survey revealed that EU and USA required expenditure from 10 percent to 60 percent of sales revenue to comply with their standards and regulation which is 10 percent to 30 percent for Japan, China and South East Asia.

Change in the expenditure on standard and regulation compliance parameters are taken as preprocessing, processing, upgrade technology, improving quality standards, HACCP Training, Sanitation facilities, Administration and Management. The benefits of compliance with standards parameters are taken as cost saving and efficient parameters and it includes. HACCP certification programme, improved consumer confidence in safety of seafood, increased prices due to compliance with standards, decrease spoilage, improved operational efficiency, green technology, market access and product differentiation.

Export Market Competitors

Analysis of the survey revealed that export competition between countries and among seafood exporters within the country. Thailand, Indonesia, Vietnam and China are the major export market competitors. Survey also specified that price competition among the seafood exporters resulted in low profit margin. Tight competition among the seafood exporters resulted in an increase in raw materials price and emergence of more intermediaries in the supply chain.

Barriers to Entry in Export

Analysis of the survey shows that the main barriers to entry in export are non-tariff barriers and anti-dumping duty imposed by US on frozen shrimp. The element of risk is higher as fish is considered as a most perishable commodity and therefore international quality assurance standards are to be ensured. Rejected seafood products are either destroyed by the importing country or diverted to new destination. Information on change of regulations by importing countries was being not informed to seafood exporting units on time. China demands additional advance of finance without documentation and it creates problems to exporters and it is another type of barrier to entry in export.

High restrictions and regulation of anti-dumping duty imposed by US has changed the attitude of the seafood exporters to export squid instead of frozen shrimp. Machine break down, container strike, scarcity of raw materials, demand of brand development, competitive price trend are other reasons of barriers to entry in export. Finance is another big concern to export, as every business needs money to roll, to invest at a time of crisis of rejections and detention. Economic recessions that have not hit adversely in Spain and Italy reflect the product differentiation and destination change. Dollar Exchange value fluctuation is another reason for profit and loss.

Extent of Non-Tariff Barrier

Saqib and Taneja (2005) found that the incidence of non-tariff measures on India's exports to ASEAN and Sri Lanka has increased. Survey findings explain the extent of non-tariff barriers faced by seafood export processing industry and examines the type of foreign affiliation. About 60 percent of the survey units accepted that the changing trade barriers in the exporting countries affect the export performance which is a real challenge for profit making and running a meaningful business. The Government agencies in importing countries takes more time to approve establishments, and the continual changes in policies delay the export permit licenses. Quality issues in European Union reduced the quantity exported to such countries. Analysis of the survey indicated that 40 percent of the surveyed units are not affected by the changing trade barriers as they are strived to cop up with the situation and succeed. They certainly are alert and have always find solutions to their problems.

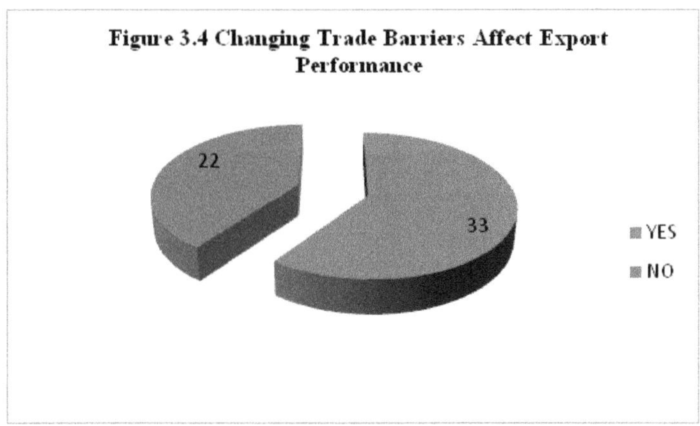

Source: Survey Data

Table No.3.12 and Figure No. 3.5 depicts the comments regarding the non-tariff barriers faced by seafood export processing firm with or without foreign affiliation.

Table No. 3.12: Non-Tariff Barriers with or without Foreign Affiliation

Remarks	With Foreign Affiliation	Without Foreign Affiliation
Facing some type of Barrier	1	26
Do not face any Barrier	10	21
Not Applicable	44	8
Total	55	55

Source: Survey Data

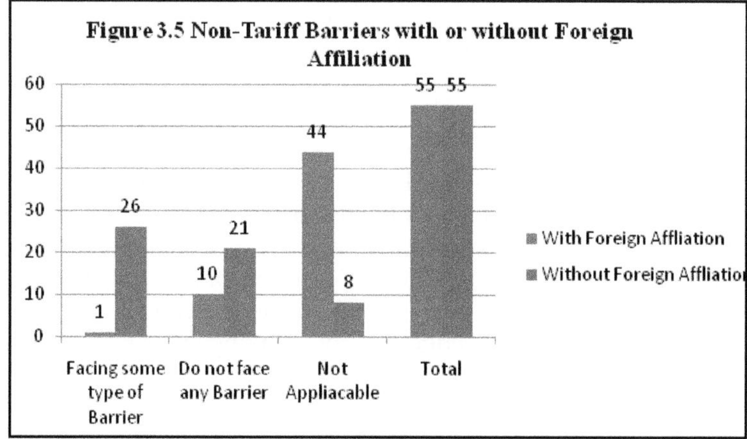

Source: Survey Data

The survey observed that there are few firms out of the surveyed units with foreign affiliation. 16 firms in the sample had some type of foreign affiliation. 3 foreign affiliated firms had joint ventures. 6 firms had a technology transfer with a foreign firm. Other type of foreign affiliation includes integrated Trademarks, Direct Supply of Raw Material, Training for Employees and Processing Plant Investment.

Table No. 3.13: Type of Foreign Affiliation of Exporting Firms

Type of Foreign Affiliation	No. of Units
Fully Foreign Owned/wholly owned	1
Subsidiary abroad	1
Joint Venture	3
Trademarks	1
Direct Supply of Raw Material	2
Technology Transfer	6
Training for Employees	13
Processing Plant Investment	4

Source: Survey Data

Analysis of the 17 surveyed units revealed that market share of post anti-dumping scenario has decreased, 2 survey units remarked it as increased while 36 survey units indicated that there is no change in the market share, it is presented in the Figure No. 3.6

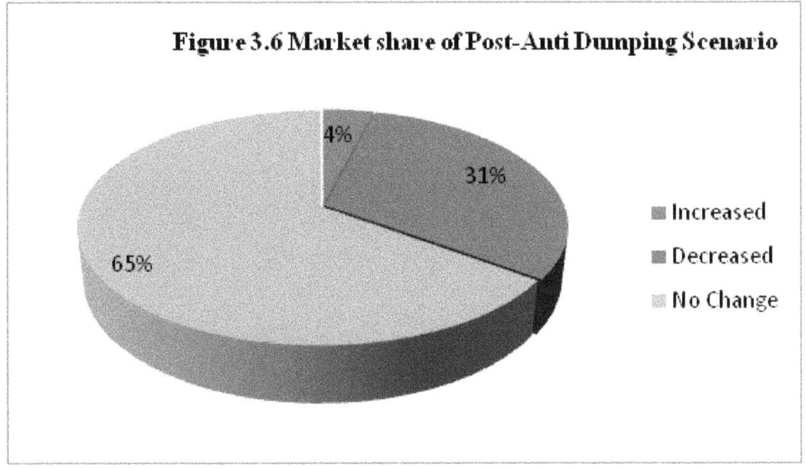

Source: Survey data

In order to solve the market share problem, Seafood exporters should comply with USA importing regulations or change seafood export products i.e., product differentiation. Survey analysis showed that instead of shrimp, squid is exported due to anti-dumping issues.

Table No 3.14: Level of Trade Restrictiveness of Non-tariff Measures

Country	Not at All Restrictive	Moderately Restrictive	Extremely Restrictive
European Union	3 (5 %)	7 (13 %)	45 (82 %)
USA	2 (4%)	29 (52%)	24 (44 %)
Japan	6 (11 %)	10 (18 %)	39 (71 %)
China	30 (55 %)	23 (41%)	2 (4%)
South East Asia	37 (67 %)	18 (33 %)	NIL
Others(Specify)	41 (75%)	13 (23%)	1 (2%)

Source: Survey Data

Another solution to the problem in seafood export to US is to eliminate frozen shrimp export permanently and export only ready-to-eat products to US which could fetch high profit margin than the traditional frozen shrimp export. Another is to divert products to other better markets. Survey analysis showed that by reducing cost of production and increasing the market access and also by increasing the finished product's price, exporters can overcome the duties imposed by USA.

Analysis of the survey showed that the level of trade restrictiveness of Non-tariff measures is presented in Table No.3.14. The Survey revealed that 84 percent of the surveyed units stated trade to EU countries is extremely restrictive. About 52 percent of the surveyed units pointed out that trade to US is moderately restrictive. Another 55 percent of the seafood exporters revealed that China's trade is not restrictive. About 67 percent indicated that South East Asian trades are not restrictive. Gulf and Iran are not at all restrictive.

4

ANTI-DUMPING MEASURES AND MARINE PRODUCT EXPORTS

Globalization pave the way to widen the international competitive advantage and involves fundamental structural shift, with emerging markets by rapidly catching up to advanced economies. It is a major source of foreign exchange earnings for the developing countries. India was the largest exporter of shrimps to the US, which stood at 94,000 tonnes in year and about half of India's overall shrimp exports were valued at $1 billion. India is the eighth-largest exporter of food items to US. Between 2005 and 2010, exports to US had been hit by the imposition of anti-dumping duty (AD) and the imposition of countervailing duty (CVD). After the duties came into effect, the number of Indian exporters to the United States declined in a significant way from 280 in 2005 to just 68 in 2009. India's exports to the United States gradually picked up after the US Department of Commerce's (USDOC) decision that "none of the seven countries including India should pay duties for their shrimp exports to US". Currently we have 192 shrimp exporters to USA. This chapter envisages the historical aspects of anti-dumping procedures, highlights the ups and downs of shrimp export trends during 2004 to 2014, shrimp productivity, and the challenges of US anti-dumping duty on Indian shrimp exports.

India's seafood industry has become one of the leading suppliers of quality seafood to all the major markets of the world. India is the second largest aquaculture producer in the world, the 2nd largest exporter of shrimps to Europe, the 3rd largest exporter of shrimps to Japan and 4th largest exporter of shrimps to US. The Major markets for Indian seafood export during 2012-13 have been South East Asia (23.1 percent), European Union (22.2 percent), USA (21.4 percent) and Japan (10.7 percent).

Indian marine products supply scenario is dependent on production cycle, climatic changes, market preferences etc. Our traditional markets like EU, Japan and USA are not faring well. However, exports to USA have increased significantly in the last two years particularly of shrimp. Indian seafood exports to Canada, Africa and Russia have shown significant growth. The US was the largest market for Indian shrimps in dollar terms. It imported 51 percent of the exports. This was followed by 16 percent to Southeast Asia, 16 percent to the European Union (EU) and five percent to Japan. According to India's Marine Export Development Authority (MPEDA), overall marine product exports reached an all-time high in fiscal year 2013-14, with 62.12 percent of that value coming from its burgeoning shrimp sector. Global shortage of farmed shrimp continues to boost India's seafood export topping USD 5 billion (EUR 4.24 billion), prospects with Vannamei shrimp bringing in a significant part of the revenue. Global shortage for shrimps due to spread of disease in farms of South East Asia has proved to be a boost to the Indian seafood industry. Indian aquaculture farms are breeding more shrimps to meet the demand, encouraged by good prices for shrimps in the global market. The shrimp farms in South East Asia are gradually recovering after the early mortality syndrome (EMS) disease attack.

Use of Anti-dumping measure as a trade protection tool has increased phenomenally in recent years (Nandana, 2005). The term dumping is used intentionally to signify the practice of price discrimination in international trade. Economists have defined dumping as transnational price discrimination or transnational sale below costs, where prices vary between national and International markets. Dumping is said to occur when the goods are exported by country to another country a price lower than its normal value. This is an unfair trade practice which can have a distortive effect on international trade. Anti-dumping is a measure to rectify the situation arising out of the dumping of goods and its trade distortive effect. Thus the purpose of anti-dumping duty is to rectify the trade and re-establish fair trade. The use of anti-dumping measures as an instrument of fair competition is permitted by WTO. In fact anti-dumping is an instrument for ensuring fair trade and is not a measure of protection per se for the domestic industry. It provides relief to the domestic industry against the injury caused by dumping. Antidumping laws are the world's most dominant form of trade protectionism, against their trading partners.

Evolution of Anti -Dumping Law

From the beginning, anti-dumping has been part of the rhetoric and mechanics of ordinary protection (Michael, 1991).Anti-dumping began as an extension of antitrust regulation, application of customs valuation procedures, and it become an effective weapon against imports. Anti-dumping rules started to develop in the early part of this century with the adoption of legislations firstly by Canada in 1904, New Zealand in 1905, Australia in 1906 and United States in 1916, which were later subjected to quite a few amendments. In 1921 the United Kingdom also enacted its first anti-dumping legislation. Antidumping provisions have been a part of U.S. trade law for over eighty years, but have been prominent only in the past two decades. The precursors to U.S. antidumping legislation emerged in the late nineteenth century from the antitrust movement and concerns about the role of unfair competition in fostering the growth of monopolies (Irwin, 2005).The Clayton Act of 1914 made price discrimination an illegal practice if it reduced competition or tended to create a monopoly. The Antidumping Act of 1916 (part of the Revenue Act of 1916) made it illegal to sell imported goods at prices substantially lower than the market value in the exporting country. This 1916 antidumping law is a criminal statute with criminal punishments.

The Antidumping Act of 1921 contains all the elements of what we now recognize as antidumping. Anti- dumping duties may be imposed if the exporter's sales price is less than the foreign market value and the dumping must be related to injury suffered by the domestic industry, then higher import duties are the appropriate remedy. The 1921 law differs obviously from the 1916 legislation. The 1916 law focuses on the intent of the exporter, whereas the 1921 law hinges on a finding of price discrimination and injury. The 1916 law is enforced in legal proceedings in the court system, whereas the 1921 law is administrated by executive agencies. In the 1916 law, dumping is related to some vague notion of predatory pricing, but in the 1921 law dumping occurs simply if foreign firms charge lower prices on products sold in the United States than in their home market, regardless of whether predation is an issue. The remedy in the 1916 law is fines and possible imprisonment, whereas the remedy in the 1921 law is higher import duties if injury to domestic producers is found. Table No. 4.1 shows the US administrative responsibilities of anti-dumping policy.

Table No. 4.1: US Administrative Responsibilities in Antidumping Policy

Period	Dumping Determination	Injury Determination
1921–1954	Treasury Department	Treasury Department
1954–1979	Treasury Department	Tariff Commission
1979–present	Commerce Department	International Trade Commission

Source: https://www.imf.org/external/pubs/ft/wp/2005/wp0531.pdf

Note: The Tariff Commission was re-named the International Trade Commission in 1974

This Act forms the basis for America's current antidumping law. The Antidumping Act was incorporated into the 1930 Tariff Act, and later amended by the 1979 Trade Act, the 1984 Trade Act, and the 1988 Trade Act. These revisions made it easier for American companies to initiate dumping complaints, and for the U.S. government to restrict foreign imports that are sold at lower prices than similar American products. Antidumping law also became a potent weapon for protectionists in other countries in the 1980s. Between 1980 and 1990 the United States, Australia, Canada, and the European Union were responsible for bringing 95 percent of all antidumping cases worldwide. That figure dropped to 80 percent between 1985 and 1992, suggesting an increase in other countries' use of antidumping law as a weapon. Unsurprisingly, the Economist pointed out that in 1988 the anti-dumping suits are emerging as the chemical weapons of the world's trade wars. Prusa and Skeath (2001) pointed out that most of the nations that adopted

AD provisions were not unaware of its technicalities and interpretations. Tharakan, and Waelbroeck(1994) envisaged that those who are interested in restraining the misuse of the antidumping provisions should concentrate their attention on the injury determination mechanism. Ved Prakash (2004) in his perception analysis have given mixed response regarding the efficacy of anti-dumping mechanism and mentioned that efficacy of anti-dumping system is eroded by circumvention.

Anti-Dumping Law in GATT and WTO

From 1948 to 1994, the General Agreement on Tariffs and Trade (GATT), facilitated to establish a strong and prosperous multilateral trading system that became more and more liberal through rounds. In the early years, GATT trade rounds with 23 contracting parties including India, concentrated on reducing tariffs alone and had hasten of high rates of world trade growth during the 1950s and 1960s, at around eight percent a year on an average. Then, the Kennedy Round in the mid-sixties with 62 countries brought about GATT Anti-Dumping Agreement and a section on development. The Kennedy Round of the General Agreement on Tariffs and Trade (GATT), held between 1963 and 1967, resulted in the GATT Antidumping Code. This code sets out guidelines under which countries may act against foreign firms that practice predatory pricing resulting in material injury to an industry based in the importing country. GATT itself does not establish specific definitions of what constitutes dumping or act against countries that dump; it simply creates the guidelines on which countries can adopt their own laws to prevent dumping.

The Tokyo Round with 102 countries, during the seventies was the first major attempt to framework agreements on non-tariff measures. The Antidumping Code was amended during the Tokyo Round of GATT, held between 1973 to 1979. The Tokyo Round agreement mandates that all antidumping investigations be reported immediately to GATT and that a semi-annual report on antidumping cases is forwarded by the signature countries to the GATT Secretariat. The eighth, the Uruguay Round of 1986-94 with 123 countries, was the last and most extensive of all, which formulate tariffs, non-tariff measures, rules, services, intellectual property, dispute settlement, textiles, agriculture and the creation of the WTO in 1995. The main focus of WTO is to improve the welfare of the people of the member countries. GATT is now the WTO's principal rule-book for trade in goods. WTO members are allowed to apply "anti-dumping" measures on imports of a product where the exporting company exports the product at a price lower than the price it normally charges in its home market and the dumped imports cause or threaten to cause injury to the domestic industry. The Committee on Anti-Dumping Practices provides WTO members with the opportunity to discuss any matters relating to the Anti-Dumping Agreement. The WTO's Committee on Anti-Dumping Practices reviewed new legislative notifications, semi-annual reports on anti-dumping actions and ad hoc notifications of preliminary and final actions taken by WTO members (WTO, 2014).The Subsidies and Countervailing Measures (SCM) Agreement regulates WTO members' use of subsidies and of countervailing measures on subsidized imports of a product found to be injuring domestic producers of that product. The SCM Committee serves as a forum for members to discuss the implementation of the SCM Agreement and any matters arising there under.

The various issues which affect seafood trade are the tariff barriers and non-tariff barriers. The tariff barriers are basically about barriers to trade as a result of tariff rates for imports or exports. The NTBs cover a wide spectrum and includes SPS, TBT, AD and Subsidies and Countervailing measures. Anti-dumping and countervailing measures are permitted to be taken by the WTO agreements in specified situations to protect the domestic industry from serious injury arising from dumped or subsidized imports. Non-Tariff Barriers (NTBs) have surged as a mechanism to protect domestic industries (Reyes, 2010). WTO is facilitating free movement of goods across borders; there still remain several ways in which protectionist lobbies continues to block free trade. Antidumping is one such instrument that has become popular for creating barriers to exports. The WTO agreement on AD stipulates the government to show that dumping is taking place and calculate its extent of price differentials and establish that it is causing injury or threatening to do so. The agreement has detailed procedures set out on initiation and investigation of AD cases.

Implementation of Anti-dumping duty agreement provides for the right of contracting parties to apply anti-dumping measures against imports of a product at an export price below its normal value. That is usually the price of the product in the domestic market of the exporting country or less than the cost of production, if such dumped imports cause injury to a domestic industry in the territory of the importing country. The Doha Declaration set forth negotiations" aimed at clarifying and improving disciplines" under the WTO Agreements on Subsidies and Countervailing Measures also known as the Anti-dumping Agreement. Although anti-dumping provisions were heavily pushed in Doha by developing nations who are frustrated with US attempts to block imports of steel and textiles, small producers in many nations will be the victims of stronger WTO rules.

Anti dumping Law in India & Trends in AD Use

India has been a member of GATT since 8 July 1948 and a WTO member since 1 January 1995.The National Law on Anti-dumping in India has been in place since 1985. However, the first Anti-dumping investigation in India was initiated in 1992 and since then the "Designated Authority" in the Department of Commerce has been handling AD cases. Anti dumping and anti subsidies & countervailing measures in India are administered by the Directorate General of Anti Dumping and Allied Duties (DGAD came into existence on April 1998) functioning in the Dept. of Commerce in the Ministry of Commerce and Industry and the same is headed by the "Designated Authority". The Designated Authority's function, however, is only to conduct the anti dumping, anti subsidy and countervailing duty investigation and make recommendation to the Government for imposition of anti dumping or anti subsidy measures. Such duty is finally imposed or levied by a Notification of the Ministry of Finance. Thus, while the Department of Commerce recommends the anti-dumping duty, it is the Ministry of Finance, which levies such duty. Since 1999 India has been one of the leading users as well as victims of the anti-dumping measures.

Table 4.2: Cases under Investigations and duty imposed by India so far against all countries (as on 30.6.2014)

Sl. No.	Country	No. of Initiations	Duty Imposed
1	China P R	166	134
2	E U	80	64
3	Korea RP	54	41
4	Chinese Taipei	52	42
5	Thailand	37	28
6	USA	37	28
7	Japan	34	29
8	Singapore	24	19
9	Malaysia	22	17
10	Russia	22	14
11	Others	162	119
	Total	690	535

Source:http://commerce.nic.in/traderemedies/Data_Anti_dumping_investigations.pdf?id=25 accessed on 1st, November 2014.

Table 4.2 depicts that India has initiated more anti-dumping actions against developing countries rather than developed countries. It clearly reveals that nearly one fourth of the total 690 actions have been initiated against China. India accounts for nearly 24 percent of the anti-dumping cases initiated against China. Others includes the following countries like Australia, Bangladesh, Belarus, Brazil, Bulgaria, Canada, Egypt, Hong Kong, Indonesia, Iran, Israel, Kenya, Mexico, Nepal, New Zealand, Norway, Oman, Pakistan, Philippines, Qatar, and Vietnam. The major product categories on which anti-dumping duty has been levied are chemicals & petrochemicals, pharmaceuticals, fibers / yarns, steel and other metals and consumer goods. Continuous use of anti-dumping action may adversely affect the small scale and export industries by raising the cost. Government has to have strong monitoring mechanism for studying the effect of dumping on small industries that are not in a position to seek protection in the form of anti dumping action (Narayanan and Natarajan, 2008).

India's share in global antidumping action has increased both as target and as an initiator. Indian exporters of steel, linen, graphite and seafood have borne the brunt of anti-dumping. European Union, USA, South Africa, Argentina and Brazil accounts for 61 percent of the anti-dumping cases initiated against Indian Exports. The European Union Continues to be on the top of the list with 34 initiations out of total 181 initiations against India .USA follows the list with 26 initiations, followed by South Africa with 22 Initiations, Brazil accounts for 17 initiations, against India as shown in Table No 4.3

Table No 4.3: Country-Wise AD Initiations: Against India From01/01/1995 to 31/12/2014

Slno	Exporting Country	Number of AD Initiation Against India
1	Argentina	12
2	Australia	4
3	Brazil	17
4	Canada	6
5	China	7
6	Egypt	9
7	European Union	34
8	Indonesia	13
9	Israel	1
10	Korea, Republic of	5
11	Malaysia	1
12	Mexico	1
13	Pakistan	1
14	Peru	2
15	Poland	1
16	Russian Federation	1
17	South Africa	22
18	Taipei, Chinese	2
19	Thailand	2
20	Trinidad and Tobago	2
21	Turkey	12
22	United States	26
	Total	181

Source: WTO, Anti-dumping Database, Geneva.

AD initiations against India accounts for 3.912 of all AD initiations between 1995-2014 and is shown in Table No.4.4 Year-wise distribution of AD initiations against India has been shown in the Table No.4.5. Antidumping cases against India may have serious consequence for Indian exports because they are the major trading partners of India.

Table No.4.4: Initiation of Anti-Dumping Cases from: 01/01/1995 to 31/12/2014

Countries AD Initiation	No. of Cases	Percentage Share
India	181	3.912
China	1022	22.09
USA	251	5.425
European Union	104	2.248
Japan	185	3.998
Korea, Republic of	341	7.37
Malaysia	119	2.572
Taipei, Chinese	258	5.576
World Total	4627	100

Source: WTO, Anti-dumping Database, Geneva.

Table No.4.5: AD Initiations against India from 1/01/95 to 31/06/2014

Year	Exporting Country India	World Total
1995	3	157
1996	11	226
1997	8	246
1998	13	264
1999	13	359
2000	10	296
2001	12	372
2002	16	311
2003	14	234
2004	8	220
2005	14	200
2006	6	203
2007	4	165
2008	6	218
2009	7	217
2010	4	173
2011	7	165
2012	10	208
2013	11	287
2014	4	106
Total	181	4627

Source: WTO, Anti-dumping Database Geneva.

From the Table No.4.6 and Figure No.4.1 it is observed that the most affected industries of anti-dumping duty are metal, chemical, plastic, textiles, machinery and equipment, agriculture and food. Product wise analysis of cases against Indian exports indicates that highest number of anti-dumping cases continues to be on the base metals account for 30 percent of the total cases. This followed the global trend also, so far as products most targeted by AD measures are concerned. This is followed by chemicals and allied products which accounts for about 24 percent of all AD measures against India.

Table No.4.6: Sectoral Distribution of AD Initiations against India From 01/01/1995 to 31/12/2014

HS Section Name	India
I Live animals and products	1
II Vegetable products	3
VI Products of the chemical and allied industries	43
VII Resins, plastics and articles; rubber and articles	30
X Paper, paperboard and articles	3
XI Textiles and articles	23
XII Footwear, headgear; feathers, artif. flowers, fans	2
XIII Articles of stone, plaster; ceramic prod.; glass	5
XV Base metals and articles	55
XVI Machinery and electrical equipment	13
XVIII Instruments, clocks, recorders and reproducers	1
XX Miscellaneous manufactured articles	2
Total	181

Source: WTO, Anti-dumping Database, Geneva.

Till 1991, India had a highly restrictive trade regime. In order to attain one of the long term objectives of self reliance our country has followed a high level of protection through import controls and high duties. After the introduction of new economic regime, India opened the economy to market forces. The first anti-dumping case was initiated in India in 1992. Until 1997, the number of AD initiations increased slowly. Since 1997-98 however, there has been a rapid increase in these cases. Recently India is one of the most active users of the AD tool. During 2001-2003 Indonesia ranks second with 26 safeguard initiations for the period. India's share in the world trade was only 0.7 percent but its share in AD initiations was over 20 percent. In 1997 India also started using safeguard measures. The use of contingent protection in India is mainly anti-dumping and safeguards measures.

India continues to be on the top of the list with 39 safeguard initiations out of the world total of 295 initiations during the period 1995 to 2014.

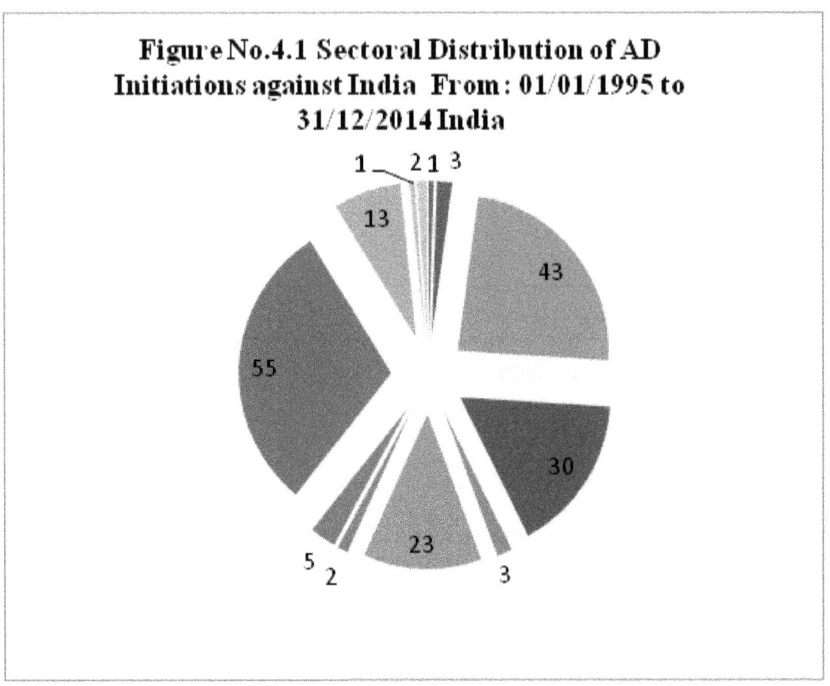

Source: WTO, Anti-dumping Database. Geneva

Import and export trends of Shrimp

The top ten shrimp importing markets in the order of ranking were the EU, the USA, Japan, Vietnam, the Republic of Korea, China, Hong Kong SAR, Mexico, Canada and Australia. These markets bought 1.3 million tonnes of shrimp during 2014, which is 8 percent or 100000 tonnes more than in the same period a year ago. However, there were negative trends in Japan (-19 percent), Hong Kong SAR (-10 percent) and Canada (-5 percent). In export trade, India, Ecuador, China, Indonesia, Vietnam and Thailand were the top suppliers. Shrimp exports from India increased by 47 percent in quantity and 71 percent in value during 2014, whereas the growth rates for Ecuador were 35 percent in quantity and 57 percent in value. Exports from China and Thailand declined by 15 percent and 31 percent respectively (Globefish, 2015). The US is the main international market for seafood from India. Its purchases represented 21 percent of total shipments, valued at INR 188.56 billion (USD 3,013.3 million) in the year 2012-13.

Table No. 4.7 gives the regression results of shrimp export value from India to US. The following are the equations used to analyse regression result of Shrimp export value from India to US with anti-dumping duty taken as a dummy variable.

ISEV= f(Puv, TSEV, INSE, T,D)

where ISEV = Indian Shrimp Export Value

Puv =Average Unit value Realization (Rs. / Kg)

TSEV= Thailand Shrimp Export Value

INSE = Indonesian Shrimp Export Value

T= Technological Changes

D = Anti-dumping duty taken as dummy variable

The regression estimates for Indian marine exports show that the variables such as Thailand shrimp export value to USA(TEUS), Dummy Variable (d) are found to be statistically significant when compared to Indian Shrimp Export to USA (Value in Crores,IUS). The R^2 and adjusted R^2 value are also high. Regression estimates show that WTO rules and regulations and Anti-dumping duty on frozen shrimp by USA significantly influence Indian shrimp export to US.

Table No. 4.8 shows regression results of the relationship between marine export quantity from India and fish production. The following is the functional form used to analyse regression result of export quantity from India. The independent variable i.e. Unit Price, US Shrimp, Total Fish Production, technological changes and Anti-dumping duty taken as a dummy variable.

EQ= f (U, EUS, FP, T, D)

Where

EQ = Export Quantity.

U = Average Unit value Realization (Rs. / Kg).

EUS = Shrimp Export USA

FP = Total Fish Production from India.

T = Technological changes.

D = Dummy Variable as Anti-Dumping Variable

Table no 4.8 regression results reveals that the variables such as Total Marine Export Quantity (EQ), Average Unit Price/kg (AUP), Export to US (EUS), Fish Production (QFP), time period (t), dummy Variable (d) are statistically significant. Fish production quantity (QFP) is having an inverse relationship with Total Quantity Export (EI).

U S Anti-Dumping Measures on Indian Frozen Shrimp

Anti-Dumping and Countervailing duty laws are administered jointly by the U.S International Trade Commission and the U.S. Department of Commerce. AD and CVD Operations conducts investigations in response to petitions received by the Department from domestic industries or labour unions. AD and CVD Operations also conducts subsequent proceedings known as administrative reviews in which importers' actual duty liability is assessed. On 31st December 2003, the Ad Hoc Shrimp Trade Action Committee (ASTAC), an association of shrimp farmers in eight southern states of the United States viz, Alabama, Florida, Georgia, Louisiana,

Table No. 4.7: Regression Result of Shrimp Export from India

Equation No.	Dependent (Variable)	Intercept (Constant)	Independent Variable					R^2	$AdjR^2$
			Puv	TSEV	INSE	T	D		
Eqn.1	ISEV	-4245.07 (11.0143)*	31.49116 (16.54221)*					0.95	0.95
Eqn.2	ISEV	6424.02 (3.670161)*		-0.01318 (-2.64199)*				0.37	0.31
Eqn3	ISEV	580.6295 (10.53039)*			0.01161 (1.338043)***			0.12	0.57
Eqn4	ISEV	-4388 (-12.2706)*	40.96299 (14.87636)*			-56.9006 (1.89476)***		0.96	0.96
Eqn.5	ISEV	-4326.88 (-1.2092)***	40.59766 (13.90114)*			-54.2493 (-1.72805)*	-151.481 (-0.55308)	0.96	0.96
Eqn. 6	ISEV	-1246.63 (0.78424)	28.47506 (4.295339)*	-0.0056 (-1.98349)***		59.15556 (0.93174)	-300.21 (-1.19024)***	0.97	0.96
Eqn7	ISEV	-1209.13 (-0.75878)	28.13187 (4.228489)*	-0.00527 (1.84697)***	-0.00421 (-0.98028)	109.8452 (1.33975)***	-248.317 (0.96151)	0.98	0.96

Source : Computed from MPEDA Government of Kerala, Kochi., World Bank Database, 2000-2014
Note :* statistically significant at 1 percent
** Statistically significant at 5 percent level
*** Statistically significant at 10 percent level
Figures in parentheses Indicates t-Statistic Value

Table No.4.8 Regression Result of Marine Export Quantity and Fish Production

Equation No.	Dependent (Variable)	Intercept Constant	Independent Variable						R^2	Adj R^2
			U	***EUS***	***FP***	***T***	***D***	***F -Ratio***		
Eqn.1	EQ	-22042 (-1.5563)	1085.665 (13.03568)*					169.93	0.928	0.928
Eqn2	EQ	7275564 (6.327724)*		2.109421 (8.276633)*				169.928	0.85	0.83
Eqn.3	EQ	-22042 (-1.5563)			1085.66 (13.0356)*			169.92	0.93	0.92
Eqn.4	EQ	17.66975 (0.001076)	786.361 (4.859559)*	0.68442 (2.0777)**				110.60	0.944	0.952
Eqn5	EQ	-333361 (-2.06056)**	490.315 (3.20921)*	0.88320 (3.4910)*	10.15112 (3.144852)*			136.63	0.976	0.969
Eqn6	EQ	159450.8 (3.21228)*	612.4646 (6.013189)*	1.025874 (6.22709)*	-30.9114 (2.93144)**	10839.1 (3.9695)*		257.671	0.99	0.98
Eqn.7	EQ	162798.9 (3.103149)*	617.4723 (5.761762)*	1.01167 (1.01167)***	-31.5546 (-2.83626)**	11026.7 (3.81984)*	-2234.6 (00.462)	188.67	0.991	0.987

Source : Computed from MPEDA Government of Kerala, Kochi , World Bank Database, 2000-2014
Note :* statistically significant at 1 percent
** Statistically significant at 5 percent level
*** Statistically significant at 10 percent level
Figures in parentheses Indicates t-Statistic Value

Mississippi, North Carolina, South Carolina and Texas , filed an anti-dumping petition against six countries- Brazil, China, Ecuador, India, Thailand and Vietnam. The petition meeting statutory requirements, on 21 January 2004 the US Department of Commerce (DOC) announced the initiation of anti-dumping investigations against the six countries. Products covered include warm water shrimp, whether frozen or canned, wild caught (ocean harvested) or farm-raised (produced by aqua-culture), head-on or head-off, shell-on or peeled, tail-on or tail-off, deveined or not-deveined, cooked or raw, or otherwise processed in frozen or canned form.

The Department notified the International Trade Commission (ITC) of its decision on initiation. On 17 February 2004, the International Trade Commission announced its decision that there was a reasonable indication that the US shrimp industry is materially injured or threatened with material injury by imports, allegedly at less than fair value, from the six identified countries. As a result, the Department of Commerce continued with its investigations and gave its preliminary determination on 28 July 2004. The ratio of preliminary duty varies between 3.56percent and 27.49 percent for three mandatory respondents selected by the DOC. The weighted arranged rate for India is 14.2percent, and the average rate for China is 49.09 percent, for Brazil 36.91 percent, for Vietnam 16.01 percent, for Ecuador 7.3percent and for Thailand 6.39percent.Indian shrimp exports were also subjected to Enhanced Bond Requirement (EBR) from August 2004 to March 2009. The antidumping duty imposed on Indian shrimps added with the continuous bond requirement by the US importers of Indian shrimp had brought a serious setback to increase India's exports to USA and this became a major trade barrier for India.

Shrimp exports to USA dropped by 23 percent in Dollar earnings during 2006-07. From Table 4.9 it is observed that frozen shrimp continued to be the major export value item accounting for a share of 64.12percent of the total US Dollar earnings. Shrimp exports during the period increased by 31.85 percent, 99.54 percent and 78.06 percent in quantity, rupee value and US Dollar value respectively.

Table No. 4.9 : Frozen Shrimp Export April- 2012 to March- 2014

Q: Quantity in Tonnes V: Value in Crores $: USD Million

ITEM		Share %	Apr 2013-Mar 2014	Apr 2012-Mar 2013	(%) Change
FROZEN SHRIMP	Q:	31	301435	228620	31.85
	V:	64.11	19368.3	9706.36	99.54
	$:	64.12	3210.94	1803.26	78.06
	UV$:		10.65	7.89	35.05
TOTAL	Q:	100	983756	928215	5.98
	V:	100	30213.26	18856.3	60.23
	$:	100	5007.7	3511.67	42.6
	UV$:		5.09	3.78	34.55

Source: MPEDA, Government of India, Kochi

The overall export of shrimp during 2013-14 was to the tune of 3, 01,435 Metric Tonnes worth US $ 3210.94 million. USA is the largest market (95,927 Metric Tonnes) for frozen shrimps exports in quantity terms followed by European Union (73,487 Metric Tonnes), South East Asia (52, Metric Tonnes) and Japan (28,719 Metric Tonnes).The contribution of cultured shrimp to the total shrimp export is 73.31percent in terms of US $. The export of cultured shrimp has shown tremendous growth of 36.71percent in quantity and 92.29percent in dollar terms. In 2013-14, the export of Vannamei has shown tremendous growth and increased to 1, 75,071 Metric Tonnes from 91,171 Metric Tonnes and US $ 1,994.27 million from 731.01 million compared to 2012-13. The export of Vannamei recorded a growth of 92.03 percent in quantity and 172.81percent in dollar terms. 44.59 percent of total Vannamei shrimp was exported to USA followed by 17.07 percent to EU, 16.54percent to South East Asian countries and 4.01 percent to Japan in terms of US $. Export of Black Tiger shrimp reduced from US $521.33 million to 435.79 million and 61,177 Metric Tonnes to 34,133 Metric Tonnes when compared to last year.

The details of frozen shrimp exports are given in Table No.4.10 USA had a market share of 32.18 percent of shrimp exports in terms of value in 2004-05, which reduced to 17.53 percent in 2009-10. Exports to USA have decreased in Quantity terms, Value in Rupee terms and Value in Dollar terms from 2004-05 to 2009-10.

Figure 4.2 clearly shows that exports to USA were showing a negative growth and this severe decrease in exports was basically due to the imposition of anti-dumping duty by USA. Between 2005 and 2010, exports to US had been affecting by the imposition of anti-dumping duty and the Imposition of countervailing duty (CVD) worsened the Indian shrimp exporters and diversifies the market scenario. From 2010-11 onwards there was a sharp increase in the market share percentage. Aquaculture production in Thailand fell about 50 percent due to the outbreak of a disease called the Early Mortality Syndrome (EMS).

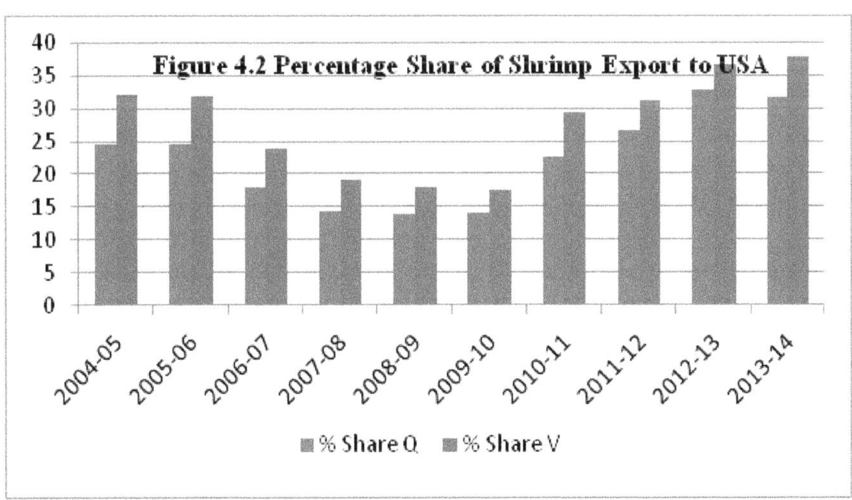

Source: MPEDA, Government of India, Kochi.

The premium quality of Indian shrimps attracted importers in the global market and India is now the eighth-largest exporter of food items to the US. The Indian seafood sector made various efforts to comply with US FDA import requirements and we have no major issues on the inspection procedures followed by FDA for seafood imports.

The Southern Shrimp Alliance has compiled refusal information for shrimp products since 2002 and is shown in the Figure No. 4.3. The FDA's 82 refusals of shrimp entry lines in the first two months of 2015 is higher than the total number of entry lines of shrimp refused for reasons related to veterinary drug residues in each of the following ten years: 2002, 2003, 2004, 2005, 2006, 2008, 2009, 2010, 2012, and 2013.

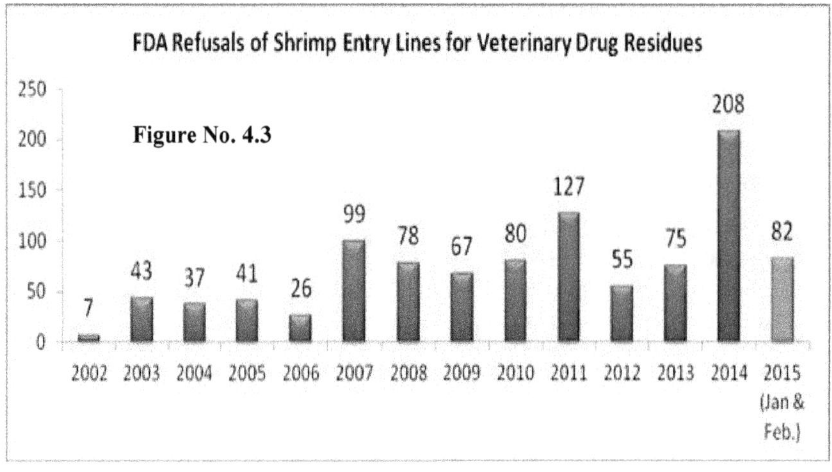

Source: MPEDA, Government of India, Kochi.

Source:http://www.shrimpalliance.com/fda-is-cracking-down-on-contaminated-shrimp- imports-update/

In fact, the 82 entry line refusals in the first two months of this year already exceeds the FDA's annual average of 72.5 entry line refusals in the prior thirteen years. The FDA's reporting for the month of February continues to provide evidence of a crackdown on shrimp imports contaminated with banned antibiotics. The FDA's reporting of refusals clearly demonstrates that problems with banned antibiotics in shrimp aquaculture have been limited to a small number of countries. In particular, of the 1,025 entry lines of shrimp refused for reasons related to veterinary drug residues since 2002, over 90percent (928) were shipped from just five countries: China (409), Malaysia (244), Vietnam (175), India (73), and Indonesia (27) and is shown in Figure 4.4.Nearly fourteen years of data demonstrate that there are a large number of sources of farmed shrimp available that do not raise significant concerns regarding the use of banned antibiotics. Yet the relentless pursuit of cheap shrimp appears to be more important.

Table No.4.10: Percentage share of Shrimp Exports to USA compared to Overall Exports 2004-2014, Q: Quantity in Metric Tonnes. V: Value in Rs.Crore, $: US Dollar Million.

Market		2004-05	Z	2006-07	2007-08	2008-09	2009-10	2010-11	2011-12	2012-13	2013-14
USA	Q	34020	35745	24697	19531	17499	18383	34243	50571	75415	95927
	V	1358.4	1362.7	1080.8	753.51	683.31	733.42	1688.5	2556	3573.8	7344.1
	$	301.55	310	237.81	187.27	152.13	154.6	371.75	548.54	663.14	1219.3
Total	Q	138085	145180	137397	136223	126042	130553	151465	189125	228620	301435
	V	4220.7	4271.5	4506.1	3941.6	3779.9	4182.4	5718.1	8175.3	9706.4	19368
	$	938.41	970.43	997.65	980.62	839.3	883.03	1261.8	1741.2	1803.3	3210.9
% Share	Q	24.63	24.62	17.97	14.33	13.88	14.08	22.6	26.73	32.98	31.82
	V	32.18	31.9	23.98	19.11	18.07	17.53	29.52	31.26	36.8	37.91

Source: MPEDA, Government of India, Kochi

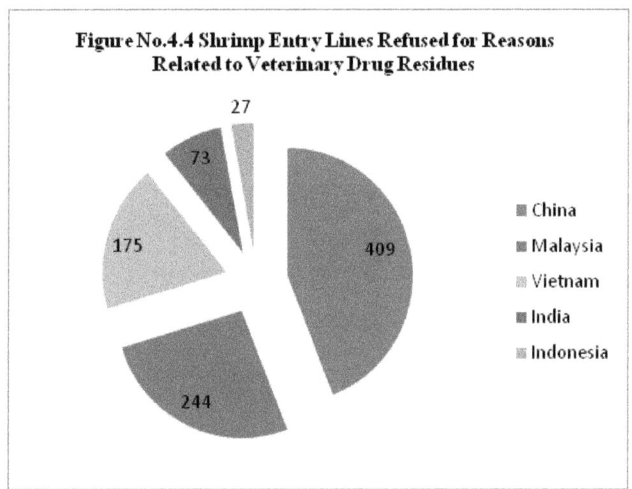

Source:http://www.shrimpalliance.com/fda-is-cracking-down-on-contaminated-shrimp-imports-update/

Production Scenario of Indian Shrimp

Shrimps are known as the pink gold of the sea, and large scale attraction of farmers towards shrimp farming has been referred to as Pink Gold. It is a universally accepted fact that shrimp has high taste factor, high unit value and have an accelerating demand in both domestic as well as international market which upholds its price. Aquaculture shrimps have been the primary contributors to the seafood export, contributing 47 percent of the total seafood export sector during the financial year 2013-14. Andhra Pradesh shrimp aquaculture exports alone contributed 27 percent of the total exports. Shrimp exports were $3.2 billion in 2013-14 which was 64 percent of the total exports of which aquaculture alone was worth $2.3 billion thus contributing 47 percent of total seafood exports. Shrimp exports from capture fisheries contributed only 27 percent of the total exports with a value of $ 900 million. The main exports were Vannamei shrimp which grew to 75,071 tonnes from 91,171 tonnes and recorded a growth of 92 percent in quantity terms and 173 percent in dollar terms when compared with the year 2012-2013.

Aquaculture over recent years has not only led to substantial socio-economic benefits such as increased nutritional levels, income, employment and foreign exchange, but has also brought vast un-utilized and under-utilized land and water resources under culture. The early 1990s witnessed a spectacular rise in farmed shrimp production with an increase from 35 500 tonnes in 1991–1992 to 82 850 tonnes in 1994–1995. Furthermore, the sector took almost 4–5 years to revive following the damage inflicted by the white spot syndrome. A cautious approach and the adoption of good management practices subsequently, helped the sector to reach a record production of 270 819 tonnes during 2012–2013 from approximately 115 826 ha area under production.

Aquaculture resources in India include 2.36 million ha of ponds and tanks, 0.798 million ha of flood plain lakes/derelict waters plus in addition 195 210 km of rivers and canals, 2.907 million ha of reservoirs and that can be utilized for aquaculture purposes. Ponds and tanks are the prime resources for freshwater aquaculture; however, only about 40 percent of the available area is used for aquaculture currently. The brackish water aquaculture sector is mainly supported by shrimp production, as well as, the giant tiger prawn which is responsible for the bulk of production followed by the recently introduced white leg shrimp, Penaeus Vannamei, the shrimp culture is again regaining its glory of export earner at large. The State wise brackish water area suitable for shrimp cultivation in India is around 11.91 lakhs ha of this only around 1.65 lakhs ha are under shrimp farming now. From Table No.4.11 it was found that the potential area has to be utilised for productive shrimp cultivation to accelerate shrimp export from India.

Table No.4.11: Status of Brackish water Land Resource Area and Area Developed in India

State	Potential Area (ha)	Potential Area (Percent)	Area Developed (ha)	Area Developed (Percent)
West Bengal	4,05,000	34.01	50,405	12.5
Orissa	31,600	2.7	12,877	40.8
Andhra Pradesh	1,50,000	12.6	76,687	51.12
Pondicherry	800	0.1	130	16.3
Tamil Nadu	56,000	4.7	5,286	9.4
Kerala	65,000	5.5	14,106	21.7
Karnataka	8,000	0.7	1,910	23.9
Goa	18,500	1.6	310	1.7
Maharashtra	80,000	6.7	1,281	1.6
Gujarat	3,76,000	31.6	2,271	0.6
Total	11,90,900		1,65,263	13.9

Source: MPEDA, Government of India, Kochi

From the Table No.4.12, it is observed that the productivity in India for Tiger Shrimp is only 1.32 Metric Tonnes /ha/yr, which is very low compared to the international standards. Productivity of Vannamei is comparatively high. India should take apt methods to enhance the productivity of shrimps to remain competitive in the international market. Increases in shrimp production in India can be attributed to the country's switch from Black Tiger to Vannamei species, which has proven very adaptable to the climate in India.

Table No.4.12: Shrimp Production in 2011 to 2013

	Total 2011-12			Total 2012-13		
Item	Area (ha)	Production (MT)	Production MT/ha/yr	Area (ha)	Production (MT)	Productivity MT /ha/yr
Vannamei	7 837	80717	10.3	22175	147516	6.49
Tiger Shrimp	114 370	135465	1.18	93110	123303	1.32
Total	122207	216182	1.76	115825	270819	2.33

Source: MPEDA, Government of India, Kochi.

Developmental Financial Assistance Schemes in India

In order to boost the exports of marine products from the country, Marine Products Export Development Authority (MPEDA) is operating various Developmental financial assistance schemes for the benefit of the seafood industry in the country.

The Director General of Foreign Trade (DGFT) on June 5th 2012, published the Annual Supplement of the Foreign Trade Policy (FTP) in which the Status Holder Incentive Scheme was expanded to cover more export product group including marine products. On December 28, 2012 marine products were withdrawn from the scope of SHIS.

Factors Taken for the Imposition of AD Measures

In 2001, 50 percent of global seafood exports emanated from developing countries whilst 74 percent of all imports went to OECD leaders, viz. USA, EU and Japan (Bostock et al., 2004). Huge differentials in labour costs mean that processing is also migrating to the developing world, especially to counties where liberalizing economies have encouraged the build of a competitive industrial infrastructure. The critical issue is not dumping but the anti-dumping measures imposed by the west that present a serious threat to developing country seafood exporters. The USA has been the main protagonist in this regard and so their anti-dumping procedures are perhaps the most relevant.

WTO Members can impose anti-dumping measures, if, after investigation in accordance with the agreement, a determination is made that dumping is occurring, that the domestic industry producing the like product in the importing country is suffering material injury, and that there is a causal link between the two. In addition to substantive rules governing the determination of dumping, injury, and causal link, the agreement sets forth detailed procedural rules for the initiation and conduct of investigations, the imposition of measures, and the duration and review of measures.

The dumped imports are causing or are threatening to cause material injury to the domestic industry. A Dumping margin represents by how much the fair-value price exceeds the dumped price. The dumping margin is the difference between the export price and the normal value. The Anti-dumping agreement gives rules for calculating the export price i.e. the price of the allegedly dumped product in the importing country, and the normal value that is the price of the allegedly dumped product in the exporting country.

Injury can be examined in two major areas i.e. Volume wise and the Price wise. The authority examines the volume of the dumped imports including the extent to which there has been a significant increase in the volume of the dumped imports either in absolute terms or in relation to production or consumption. The dumped imports are causing price depression or it may prevent it to increase price. The agreement sets out a series of factors concerning the volume of dumped imports, effects of dumped imports on prices, and the impact of dumped imports on the domestic industry, which must be considered in that analysis. Regarding the impact of dumped imports on the domestic industry, the agreement requires consideration of all relevant factors, including sales, market share, capacity utilization, employment levels, ability to raise capital, return on investments, cash flow, inventories and the magnitude of the margin of dumping and investments.

The agreement leaves a great deal of discretion to the investigating authorities as to how a causal link is to be established. Members have been relying on various methods for reaching a determination e.g. finding out a link between the injury and the dumped imports, as well as undertaking econometric analysis to estimate the performance of the domestic industry had there been no dumped imports. The Agreement requires that the investigating authorities also examine other known factors which may be causing injury for example imports not at dumped prices, contraction in demand or changes in the pattern of consumption, development in technology, trade restrictive practices or competition by other producers and productivity of the domestic industry. If such other factors are also found causing injury this cannot be attributed to the dumped imports.

Zeroing

Under the WTO Anti-dumping Agreement, a country can impose anti-dumping duties on an imported product if they find that the company exporting the product charges less for it in the market of the importing country than in the exporting country. To determine the antidumping margin, i.e. the difference between those two prices to determine the level of the duty to be imposed, the US DOC used a methodology known as 'zeroing'. From the period of GATT, zeroing has been an issue for debate as it tends to inflate final dumping margins by preventing negative margins from offsetting positive margins. It is a process that has already been challenged in number of previous disputes between the US and countries, such as Canada, the EU, Japan, Mexico, Ecuador and Thailand. Each time, the WTO DSB has constantly ruled against this practice, where investigators treat transactions with negative dumping margins as having margins equal to zero in determining weighted average anti-dumping margins. For example, if a good was sold for $100 in the home market and $70 in the foreign market, a 30 percent dumping margin would be applied, but if the good was sold for $130 in the foreign market, a zero value would be applied. Thus, when aggregating these translations, a dumping margin of $30 would be found whereas an average of the two transactions would have resulted in no margin and hence no dumping. The possible effects of dumping on the seafood processing industry are decline in sales, market share, profitability and employment. The antidumping investigation also has harassment effect in terms of time and financial cost to provide data as required by the regulatory authorities and for organizing legal support. For the small exporting firms from developing countries it is too expensive to take advantage of the procedure. In 2004-2007 the EU launched WTO disputes against the US for the use of a methodology called "zeroing" in the calculation of antidumping duty rate. The EU's initiative was followed by a number of countries including India, Brazil, China, Ecuador, Japan, Korea, Mexico, Thailand and Vietnam. The WTO Appellate Body has consistently condemned the practice of zeroing over the past decade as unfair. The US finally committed itself to comply with the main elements of the WTO ruling by removing zeroing in all ongoing and future cases.

Financial Adversity to US Shrimp Exporters from AD Proceedings

The US Customs and Border Protection (CBP) have withdrawn the enhanced bond requirement (EBR) with effect from 1st April 2009. However CBP does not intend to cancel the bonds and treated as normal customs bonds given during the tenure EBR from August 2004 to March 2009.The Indian IORs (Importers of Record) executed customs bonds for minimum required amount of $50,000. In August 2004, CBP introduced EBR for frozen shrimp, thereby requiring the bond among to be calculated at 10 percent of past or potential antidumping duties. Thus the EBR enhanced the bond requirement manifold.

For example:

Estimated Import Value in a Year 1	-	$10million.
Rate of Anti-dumping Duty	-	10percent
Estimated Antidumping Duty Deposit	-	$1million
EBR for Year 1	-	$1million
Normal Customs Bond	-	$0.050million
Total Bond Required	-	$1.050 million

Since antidumping duties under U.S. retrospective system take several years to be assessed, a new bond is required to be executed once the limits under an existing bond are exhausted. Thus each IOR has at any time three or four bonds executed under EBR pending with CBP.

For example:

Bond Executed As above	-	$1.050million
Estimated Import Value in Year 2	-	$12million
Rate of Antidumping Duty	-	10percent
Estimated Antidumping Duty Deposits	-	$1.2million
New EBR for Year 2	-	$1.2million
Cumulative EBR Outstanding	-	$2.250 million

For executing the bonds, the surety companies in US demanded 100 percent collateral. Therefore the IOR in above example provided collateral by way of irrevocable letter of credit in favour of Surety Company for $ 1.050 million for Year 1, $1.2 million for Year 2 and so on. The Banks provided these collateral charge annual fees to keep each collateral as valid. Thus at the end of Year 2, the IOR in above example would have paid annual fees to the banks on $ 1.050 million for Years 1 and 2, on $1.2 million for Year 2 and so on. It is estimated that the Indian IORs annually pay $1 million to their bankers to keep the collaterals valid. Each letter of credit furnished as collateral will be outstanding of a period of four and half years. The IOR is required to annually renew the collateral five times incurring the underlying costs. This is repeated in respect of each of about 4 to 5 bonds executed between August 2004 to March 2009. EBR bonds are different from normal customs bonds. EBR are special purpose bonds whereas normal customs bonds are perpetual in nature. The Indian IORs of frozen shrimp request the Commerce Ministry to take up with USTR to use its good offices to persuade CBP not to equate the bonds under EBR with the normal customs bonds and to create a mechanism immediately to cancel and discharge these bonds, so that the surety companies can discharge the collaterals for these bonds. They also request to persuade CBP to re-liquidate the entries that were erroneously liquidated at deposit rates and to immediately refund the excess of deposits over assessment rates, together with interest up to the date of refund.

The Southern Shrimp Alliance and Impact on AD

Anti-dumping legislations have gone far beyond their mandate of promoting fair trade and have emerged as vehicle of contingent protection (Krugman, 1993).Government of India should take up the zeroing issue urgently with USA Administration and if they are not agreeable to stopping zeroing in their administrative review it should be taken up with the WTO. Many exporters feel that if zeroing is stopped Anti-dumping duty will come to minimize level. The Government of India took up with the Dispute Settlement Body (DSB) of WTO that the continuous bonds on February 2008 and USA has ended the Enhanced Bonding Requirement for imports of shrimp from all countries including India w.e.f. 1st April 2009. Accordingly the enhanced bonding was reconsidered by the US from 1st April 2009. This gives great relief to the exporters. Sunset review (SSR) meeting took place in January 2010 to consider whether the antidumping duty is to be continued or done away with. The final hearing of the SSR took place on 1st February 2011 in Washington. In its notice in Federal Register dated 29th April 2011 US SD and International Trade Commission (ITC) felt that revocation of the AD would be likely to lead a continuation or recurrence of dumping and of material injury to an industry in the US within a reasonable foreseeable time.

Replacing Enhanced Bonding Requirements Effective April 1, 2009, Customs and Border Protection's (CBP) eliminated the enhanced bonding requirement on all shrimp imports in response to an unfavorable decision from the WTO on only two of the six countries subject to antidumping orders. CBP developed the enhanced bonding program, which required that parties importing shrimp are subject to antidumping duties and obtain a bond in the amount of duties expected to be paid on imports in the prior year, in response to a well-documented and persistent failure to collect all antidumping and countervailing duties-particularly on agriculture and aquaculture products. Even with the enhanced bonding requirement, CBP reported that it has been unable to collect nearly $68.9 million in antidumping duties on shrimp.

Litigating the antidumping orders every year, the Department of Commerce (Commerce) provides an opportunity, called an "administrative review," to identify increases or decreases in unfair trade since the duties were first imposed. Commerce adjusts the antidumping rates based upon the findings of a year-long investigation: more dumping results in higher antidumping rates, and less dumping results in a refund of collected duties. First administrative reviews In the first administrative review, Commerce exercised its discretion to the benefit of foreign companies that operate in multiple countries by ignoring the "multinational corporation provision" for shrimp exporters operating in Vietnam and China. The provision enacted by Congress is designed to prevent companies from shifting production from a country with high duties, like China, to a country with lower duties, like Thailand. The SSA challenged this decision because it allowed more than a dozen large producers/exporters in Thailand to evade significantly higher duties. In May 2009, the Court of International Trade (CIT) upheld Commerce's decision. Three months later, the Ad Hoc Shrimp Trade Action Committee filed a legal brief before the Court of Appeals for the Federal Circuit that challenges the CIT decision. In December of 2009, the Court of Appeals for the Federal Circuit held oral argument on the appeal. The domestic shrimp industry was represented in oral argument by Picard Kentz & Rowe LLP. The law firm, now counsel for the SSA, was formed by three former partners of Dewey & LeBoeuf that specialize in international trade, litigation, international arbitration, tax, and public policy. The firm's attorneys have extensive international trade backgrounds, including extensive work with the U.S. shrimp industry. In the second administrative reviews commerce issued final results regarding the second administrative reviews of shrimp imported from Thailand, Ecuador, India, Brazil, and Vietnam in August 2008. Incredibly, the second administrative review regarding shrimp from China was terminated because Chinese exporters claimed to have shipped no shrimp subject to the antidumping duty order between February 1, 2006 and January 31, 2007 (the second administrative review period). The SSA appealed to the CIT in a number of determinations made by Commerce in the second administrative reviews, such as Commerce's decision to ignore procedures established by the regulations and instead allow a Thai exporter to withhold necessary information without penalty. The effect of that and other Commerce decisions was to reduce the amount of duties owed from more than 15 percent to less than 3 percent. Briefing and oral argument for each of the second administrative reviews appealed was completed in 2009. A decision rejecting the domestic industry's challenges to the final results of the second administrative review of shrimp from Ecuador was rendered on October 30, 2009. At the same time, a decision rejecting a foreign exporter's challenge to the final results of the second administrative of shrimp from India was handed down by the Court on June 24, 2009. With respect to shrimp from Vietnam, on September 29, 2009 the Court ordered that Commerce further consider distinct challenges brought separately by the domestic industry and foreign exporters. Three months later, the Court ordered

Commerce to re-examine the challenges brought by foreign exporters to the final results of the second administrative review of shrimp from Thailand.

In the third administrative reviews, the final results of the third administrative reviews of the antidumping duty orders on shrimp from China, Ecuador, India, Thailand, and Vietnam were issued in 2009. No review was conducted on the antidumping duty order for shrimp from Brazil as no Brazilian shrimp was imported into the United States between February 2007 and January of 2008 (the third administrative review period). The outcomes of these administrative reviews were mixed. With regard to China, two shrimp exporters were assigned 9.08percent dumping margins, while all other Chinese exporters subject to the review were determined to be dumping at a rate of 122.81percent. Following three successive proceedings wherein at least one Ecuadorian exporter was found to have not dumped shrimp into the United States, Commerce concluded that all shrimp entering the United States during the third administrative review period from Ecuador are subject to the antidumping duty order. Although final assessment rates for Ecuadorian shrimp was established as being between 0.75 percent and 0.85 percent, these amounts represented an increase over the dumping margins calculated for Ecuadorian exporters in the second administrative review (0.64 percent). The final dumping margins calculated for Indian exporters declined again in the third administrative review, falling from an average of 10.17 percent in the investigation, to 7.22 percent in the first administrative review, to 1.69 percent in the second administrative review to 0.79percent in the third. One Indian exporter was, for the second year in a row, found not to be dumping shrimp into the United States. Similarly, Commerce concluded that five different Vietnamese exporters were not dumping shrimp into the U.S. market during the third administrative review period. All other Vietnamese exporters were determined to be dumping at margins ranging from 4.30percent to 25.76percent. Finally, the dumping margins determined by Commerce for Thai companies increased in the third administrative review period (average duty rate = 4.71 percent) over the second administrative review period (average duty rate = 3.18 percent) but remained below the dumping margin determined in the original investigation (average duty rate = 5.95 percent). The final results of the third administrative reviews of the antidumping orders on shrimp from China, Thailand, and Vietnam have been appealed to the Court of International Trade and litigation regarding these appeals will begin in 2010. In the fourth administrative reviews, administrative reviews for the fourth review period (February 2008 through January 2009) of the antidumping duty orders on shrimp from China, India, Thailand, and Vietnam were initiated in 2009. Because the antidumping duty order on shrimp from Ecuador was revoked in 2007 as the result of an adverse ruling by the World Trade Organization, no review was conducted with respect to Ecuadorian shrimp. Also, for the second successive year, no Brazilian shrimp was imported into the United States and, as such, no administrative review was initiated regarding the Brazilian antidumping duty order. Administrative litigation in the fourth administrative reviews of the antidumping orders on shrimp from China, India, Thailand, and Vietnam will continue into 2010. The final results of these proceedings are anticipated to be released sometime between July and September of 2010.

The 6th Administrative Review on Antidumping of frozen warm water shrimps was conducted for the period 01.02.2010 to 31.01.2011. M/s. Apex Exports and M/s. Falcon Marine Exports Ltd., were the mandatory respondents. After the 6th Administrative Review, Anti Dumping Duty for M/s. Apex Exports is 2.51percent and for M/s. Falcon Marine Exports Ltd., it has been reduced to 0.13percent. Antidumping duty for other companies will be 2.51percent (Minister of State of Commerce & Industry, March 4th, 2013).

In the Seventh administrative review, Commerce had found that all Thai and Vietnamese shrimp exporters participating in the proceeding, as well as two of the three individually reviewed Indian shrimp exporters, were selling shrimp into the U.S. market at fair value. In result, Commerce declined to assess antidumping duties and did not require cash deposits for future entries. However, in the preliminary results of the current i.e. Eighth administrative review, Commerce has preliminarily found that all Indian, Thai, and Vietnamese shrimp exporters participating in the proceeding sold shrimp into the U.S. market at less than fair value.

US DOC is modifying its method of calculating the weighted-average dumping margins and anti-dumping duty assessment rate. Formerly, it compares the transaction-specific export prices and average normal values to arrive at the value of dumping. However, US DOC does not offset the amount of dumping. The new method of calculating duty may lead to de-minimum duty (below 0.5 per cent), which in effect is zero anti-dumping duty on export to US.

It is inevitable that every country would be forced to sell a very small portion of its export consignment beneath fair-value price mainly under distress conditions, the practice of the US Customs to identify these specific consignments and charge anti-dumping duty on all shipments is known as zeroing. The World Trade Organization, in recent rulings, has declared zeroing as an illegal practice under the WTO guidelines as it was found violating several international and multilateral trade rules. The removal of zeroing will be welcomed by the Indian trade and is expected to strengthen the country's shrimp exports further.

Antidumping (AD) and countervailing duty (CVD) investigations are triggered by a petition filed by an interested party on behalf of an industry alleging that the industry is injured or threatened with material injury by reason of imports that are, respectively, sold in the U.S. market at less than fair value (dumped), or subsidized. In order for the industry to obtain relief, two things must happen: (1) the International Trade Administration (ITA), an agency of the Department of Commerce, must find dumping or subsidization and (2) the International Trade Commission (ITC) must find that the domestic industry is materially injured or threatened with material injury due to the dumped or subsidized imports. These agencies conduct preliminary and final investigations in a detailed administrative process with specific time lines.

The CDSOA requires CBP to distribute all duties collected pursuant to AD or CVD orders to "affected domestic producers," defined in the act as any manufacturer, producer, farmer, rancher, or worker representative (including associations of these individuals) that was a petitioner or interested party in support of a petition that resulted in an AD or CVD order, and remains in operation. Distributions under the act may be used to offset "qualifying expenditures" within the following categories that the domestic producer incurred between the issuing of an AD or CVD order and its termination. (1) Manufacturing facilities (2) equipment, (3) research and development; (4) personnel training; (5) acquisition of technology; (6) health care benefits for employees paid by the employer; (7) pension benefits for employees paid by the employer; (8) environmental equipment, training, or technology; (9) acquisition of raw materials and other inputs; or (10) working capital or other funds needed to maintain production.

The Byrd Amendment controversy is one component of a larger debate in Congress concerning the overall direction of U.S. trade policy. Although many Members acknowledge that there are benefits received by liberalizing trade flows, there is sometimes disagreement on the proper balance between these benefits and the transition costs incurred to domestic industries, firms, and workers by increased global competition. Because the added welfare from trade tends to be diffused over the population as a whole, while losses fall disproportionately on import-competing industries and regions, Members' perceptions of free market policies differ. With regard to the CDSOA, supporters maintain that the measure helps level the playing field by compensating U.S. producers adversely affected by unfair trading practices. Now that some co-complainants have assessed WTO-authorized retaliatory duties on U.S. exports, however, Congress may later face as much pressure to repeal the measure from U.S. exporters as it does from the domestic producers who benefit from the measure.

Coalition of Gulf Shrimp Industries Fact Sheet on Shrimp Subsidy Cases.

On December 28, 2012 the coalition of Gulf Shrimp industries filed petitions with the US government seeking relief from subsidized shrimp imports from seven countries. The petitions seek the imposition of countervailing duties on shrimp from China, Ecuador, India, Indonesia, Malaysia, Thailand and Vietnam. Imposing duties to counter these harmful foreign subsidies is critical to the survival of the US Shrimp industry and the way of the life of shrimpers, docks and processors across the Gulf region. Since 2009, producers in these seven countries have gained US market share by aggressively undercutting domestic prices. This underselling has suppressed and depressed the price US producers are able to get for their product, making it more and more difficult for them to cover their costs of production. As a result the small operating margin the industry enjoyed in 2009 has disappeared. The Coalition of Gulf Shrimp Industries stated that in 2011 the seven countries exported over 984 million pounds of shrimp to the United States, worth nearly $4.3 billion and accounting for 85percent of USA imports and over 75percent of domestic consumption.

Shrimp is a major export commodity in each of the seven countries; the foreign governments have set specific growth and export targets for their domestic shrimp industries as part of their national economic development plans. To meet these targets, government is spending billions of dollars on subsidies for their shrimp industries. The petitions document more than $13.5 billion in government subsidies to the aquacultures and seafood processing industries in these countries, with the shrimp industry as the primary recipient. The COGSI petitions are being investigated by the US International Trade Commission (ITC) and the Department of Commerce (DOC), with final determinations expected by the DOC on subsidy levels on August 13, 2013 and by the ITC on injury to the US shrimp industry on Sept 19th 2013. On May 29th, the DOC preliminarily determined that shrimp producers in the seven countries benefit from a wide array of subsidies including government grants, cheap loans, debt forgiveness, tax breaks and numerous export subsidies.

For example: The Government of Thailand cuts taxes for shrimp producers that commit to export their product. The Indian government provides subsidies to reduce shrimp processors ocean freight costs, with an added subsidy specifically for exports to the US. In China, the government gave subsidized loans to shrimp producers. The Government of Vietnam provides land to shrimp processors at below market rates. The Malaysian government is investing tens of millions of dollars to build vertically integrated shrimp farms and processing facilities to target world export markets. The DOC found preliminary subsidy margins ranging from 1 to 62 percent and instructed Customs to begin collecting bonds to cover the duties. The preliminary subsidy margins may be adjusted in the final determinations on August 13. The Coalition of Gulf Shrimp Industries was formed in support these petitions and to work for the long term survival of the entire Gulf shrimp industry. The domestic producers supporting the petitions account for the vast majority of domestic production, and they represent the industry across the coastal states of Alabama, Florida, Georgia, Louisiana, North Carolina, Mississippi, South Carolina and Texas.

Today the trade remedial measures have become the most favoured non tariff measure to promote the interests of the domestic industry for most of the WTO members. Imposition of duties such as anti-dumping countervailing or safeguard duties has an important bearing on the trading decisions. While these duties levied provisionally or finally definitely make the imports costlier from such sources even a mere initiation of investigation scares away the importers. Considering the proliferation of the use of trade remedial measures it is most appropriate that an assessment is made about the impact these measures are having on the Indian economy. These measures impact the price realization by domestic industry which has the opposite effect on user industry or consumer, there is likely to be significant impact on Sales volume turnover employment inventory etc. which need to be examined and quantified. It is with the objective of making an assessment of the impact of these measures that this study has been undertaken. The basic objective of the study is to evaluate the economic impact of trade remedial measures in terms of various economic parameters like domestic producers profitability trade chilling and trade diversion effects production and location pattern changes adequacy and efficacy of these measures in protecting legitimate interests of domestic industry.

Anti-dumping duty imposed by US on Indian frozen shrimp reaches a record time of 10th anniversary. The best way to evade the AD duty is product differentiation with value added products like ready to pan or ready to eat. This will surely enrich fish processing exporters to capture greater gains from trade and wider market access which will automatically develop brand name for our products. Recent records showed that around 5 percent of Kerala's shrimp exports are value added products. Rightly this is the time we have to shift our production into finished products and come out the vicious circle of raw product to finished products. These transformations surely enrich the seafood exporters into highest foreign exchange earnings as well as high status in the society. This will definitely lead to improvement in the standard of living of the fishery sector and at the same time create productive employment and accelerate the consumer demand and producers can attain economies of scale as the well known economist Paul Krugman's New Economic Theory. Our fisheries sector can transform the entire outlook of the sector into a well structure and everlasting profit making sector. Moreover the fish processing exporters become the queen among the exporters with the support of public private partnership which extends from the catch to table. U.S. antidumping law is a tangled and confusing subject because U.S. law and procedures have changed substantially over time and due to this Indian Frozen Shrimp exporters foresee a market diversification and product differentiation challenges in order to attain better exporter earnings.

5

CHALLENGES FACED BY MARINE PRODUCT SECTOR

Fish and Fishery product exports can be an engine of economic growth in India especially in Kerala. Investing in the fishery sector made a dominant icon to reduce poverty and food security. Increased fish and fishery productivity generates a higher income and creates income generating opportunities to come out from the vicious circle of poverty. In order to fill the gap between demand for and supply of fish production, developing countries have set down an ever increasing share for growing aquaculture industry to attain a rapid growth in fish production. Net exports from the developing world are projected to continue through 2020, though at a lower level than present. This is mainly due to rising domestic demand within developing countries for fish because of population growth, income growth and urbanization (Delgado et al., 2003). This chapter analyses the supply chain and problems of fish export processing industry units in Kerala. The study tries to capture the demand side constraints to cover those related to international trade such as non-tariff barriers, while supply side constraints reflect domestic challenges in seafood processing industries, including issues related to the sustainability of natural resources.

India is the top most country which is providing total, direct and indirect marine employment in the world and the second position goes to China and third to Indonesia. Fish and Fishery products are among the most traded food commodities worldwide where developing countries account for the bulk in world exports. The share of total production that is exported increased significantly from 25 percent in the mid-1970s to nearly 40 percent in 2011, reflecting the sector's growing degree of integration in the global economy (FAO, 2012). The Seafood Export Processing industry has undergone several changes. The Seafood Service Sector has subject to drastic changes in terms of the types of products in demand. Consumer preferences and tastes are strongly in favour of ready-to-cook and ready-to-eat convenience foods and eating in restaurants has become immensely popular in domestic as well as foreign market. By exporting seafood worth of US $ 5 Billion, the seafood sectors have earned a remarkable position in India's export basket. Table No. 5.1 shows the region wise details of exporters from India and Kerala. The Gross State Domestic Product of the State has increased by about 97 percent during the period from 2005-06 to 2010-11 and the share of fisheries sector in the State Domestic Product has declined from 1.81 to 1.29 percent in the same period. The share of primary sector in GSDP has also declined from 18.05 to 14.07 percent.

Table No. 5.1: Region-wise Details of Exporters

	KERALA		INDIA	
Year	Manufacturer Exporters	Merchant Exporters	Manufacturer Exporters	Merchant Exporters
2005	57	34	254	202
2006	70	96	304	374
2007	74	84	310	341
2008	82	71	313	325
2009	90	74	340	332
2010	94	69	363	345
2011	104	67	392	387
2012	111	72	442	447
2013	113	78	471	465

Source: MPEDA, Government of India, Kochi.

While marine fish production in Kerala tends to fluctuate, the inland fish production showed a sign of improvement from 1999-2000. During 2010-11, the marine fish production has decreased to 5.6 lakh tonnes from 5.70 lakh tonnes of 2009-10. Inland production sustained on increasing trend. District-wise marine fish production showed that Alappuzha contributed the highest (23.08 percent) followed by Kollam (20.35 percent) and Kozhikode (14.95 percent). During 2010-11 the share of inland fish production to the total fish production of the state was 17.78 percent. In recent year's liberalization policies, technological innovations, improvements in processing, packaging and transportation, as well as changes in distribution and marketing have further accelerated this trend, while facilitating the emergence of complex supply chains in which goods often cross national borders several times before final consumption (Sumaila et al., 2014). Table No.5.2 depicts the monthly registration progress report of exporters, manufactures, merchant exporters, processing plants, storages, ice plants, peeling sheds and conveyance.

Table No. 5. 2: Monthly Progress Of Registration as on October 2014

Entry Name	Registered as on 1st of Month
Exporters	1107
Manufacturer Exporters	489
Merchant Exporters	504
Processing Plant	470
Storages	573
Ice Plant	78
Fishing Vessel	9803
Peeling Shed	610
Conveyance	184

Source: MPEDA, Government of India, Kochi.

Harikumar & Rajendran(2007) examined the present expansion of fisheries sector in the State and found that it is much better than the past and there is much scope for modernization and diversification of the existing scenario. At the same time there is great need to introduce deep sea fishing technologies, diversification of existing fishing fleet, addition of the recent trends in fish processing industry to cope with the international standards and efforts for boosting the coastal & inland aquaculture sector and introduction of cold storages and cold chain and modernization of the fish markets. They suggest that production of value added fish product is another area where the State has high potential. This enables us to generate more income and employment opportunities. In order to avoid a major catastrophe in the rare species of fish diversity of the State into oblivion, some earnest efforts are required to conserve, preserve and propagate them.

Indian food processing industry is the fifth largest industry in the world in terms of production, consumption, and export and growth aspect. A distinct feature of the processing sector in Kerala is its dependence on the preprocessing sector which is popularly known as "peeling sheds". Table No.5.3 shows the details of peeling sheds with capacity from India and Kerala. Efforts should be taken to register all peeling sheds in any of the government agencies like MPEDA, EIA etc to have proper infrastructure facilities and working conditions in place. Steps should also be taken to include the workers in peeling sheds in the MG scheme to ensure minimum wages are paid to the workers and also to attract more women workers into this sector. By all means the peeling sector has to be revamped by providing much assistance to rejuvenate it as it plays a major role in the seafood cold chain in India (Sathyan et al., 2014).

Table No.5.3: Region-wise Details of Peeling Sheds With Capacity

Year	Kerala		India	
	No	Capacity	No	Capacity
2005	130	768.68	396	3588
2006	163	1019.78	451	4071.8
2007	231	1504.78	536	4745.5
2008	259	1734.98	576	1599.08
2009	280	1909.31	597	5403.7
2010	277	1908.37	584	5478.55
2011	276	1900.62	588	5561.17
2012	286	1958.92	605	5738.71
2013	289	1977.12	608	5965.47

Source: MPEDA, Government of India, Kochi.

Fish Processing Export Industry in Kerala

Kerala has a higher Human Development Index than all other states in India. The state has a literacy rate of 94.59 percent, the highest in India. A survey conducted in 2005 by Transparency International ranked Kerala as the least corrupt state in the country. Fisheries sector is witnessing radical changes and challenges at national and global level. Fish and fishery products will persist to be highly traded, invigorated by increasing consumption of fishery commodities, trade liberalization policies, globalization of fishery trade and technological innovations in processing, preservation, packing and transportation. The demand for fish and fishery products are steeply rising and at the same time taste and preferences of the next-generation consumption patterns are ever changing. The emerging challenges and opportunities call for a paradigm shift in the innovation and inventions of modern infrastructure and advanced techniques pave the way for structural revolution of fishery sector into over capitalization which affected all the stakeholders of the entire supply chain as well as value chain of fish and fishery trade. Fish and fishery products in Kerala are facing crisis due to stagnation in production, low capacity utilization and highest cost of production due to overcapitalization and low productivity. There are various challenges faced by Kerala fish export processing industry on account of product diversification, dynamic market access, changing quality standards, climatic change, and global pressures and changing world scenarios.

Constraints for the Seafood Export Processing Industry

Quality and food safety is the foundation of fish export processing industry. The major barriers in the external market in the post liberalization era have been in the forms of Sanitary and Phyto- sanitary measures imposed by the developed countries like the US, the EU and Japan and the US Anti-dumping duty imposed on Indian frozen shrimp. The measures have adversely affected the prospects of the industry. The strict enforcement of quality standards by the US and the EU led to hasty implementation of Codex standards and HACCP regulations, which entailed additional cost by way of both fixed and variable costs, which ranged between 10 to 20 percent of the total cost [Ouseph, 2005]. Fishing industry is facing a lot of problems. Demand side issues are Quality Issues, International Standards and Regulations, Labeling and Certification Requirements, SPS, Codex Standards and Anti-dumping duty. Supply side issues are low levels of mechanization, low productivity, low capacity utilization, varying quality safety and hygiene, inadequate infrastructure, inadequate supply and quality of raw material, inadequate access to finance, marketing, changing business cycle and Government legislation. Moreover, the major problems faced by the fish export processing industry in Kerala are finance, labour, marketing, infrastructure, waste utilization, power, electricity charges, trade expenses, tax escalation, and increase in raw material cost, wages and salaries resulting in cost of production and profit margin. The fish storage facility in Kerala is grossly inadequate compared to the potential for fish production and processing. Extensive network of refrigerated handling, transport, storage and retailing has to be put in place. Also we have to make better use of fish waste and its by-products.

Demand side Issues

There are various demand side issues faced by the fish and fishery export processing industry. Fish export processing industry faces new challenges ensuring safety and quality standards issues which continue to be an important element in the industry development. A food substance is said to unsafe when it contains any physical, chemical or biological agent that can cause adverse health effects. Such hazardous matter could include bacteria, viruses, parasites, worms that cause various kinds of food borne illness or biological toxins, chemicals contaminants like residues of fertilizers, pesticides, insecticides, industrial wastes, veterinary drug, allergens, additives and preservatives or heavy metals like copper, mercury, arsenic and tin etc. The officers of MPEDA attended training on the "EU Feed Rules and Feed Import Requirements for Third Countries "at Italy Rome from 4th to 7th December, 2012. Twenty Six Rapid Alert Notifications were received during the year 2012-13. Thirty Rapid Alert Notifications were received during the 2013-14 financial year and concerned Regional Offices and Sub Regional Offices have investigated and have furnished their reports.

Food Safety and Quality Standards

As rising demand has led to higher prices, both quality and standards have improved. Differences in prices for different species, standards and quality grades have increased giving producers and distributors a much better incentive for supplying high quality seafood. In order to enhance seafood quality control, China has currently carried out the principle of security management. The national authorities have issued serial laws, regulations and standards related to seafood quality control such as The Product Quality Law, The Food Hygiene Law, The Standardization Law and The Commodity Inspection Law, and instructed The Ministry of Agriculture, The State Technique Monitoring Bureau and The State

Import-Export Commodity Inspection Bureau to work together for seafood quality control. For those exported seafood, China primarily adapts the existing international standards or the standards and regulations by targeted countries in order to meet their demands. For example, the processing method for baked eel exported to Japan is based on the requirements by Japanese Government and the traders. To ensure the quality of exported seafood, the processing enterprises have been gradually instructed to follow the international quality management system, such as HACCP created by Food and Drug Administration (FDA) and the National Oceanic and Atmospheric Administration (NOAA), the Seafood Quality Management Criteria by Canadian Fishery and Marine Ministry and the related regulations by the European Union.

Food safety standards and more complex requirements are demanded by the top retailers. Safety, traceability, and critical rules are required for maintaining better health standards. Demand side issues are mainly quality issues related to international standards and regulations, SPS and TBS. Supply constraints includes mainly the production issues, market diversification and financial issues. HACCP system is important for maintaining food safety in fish export processing industry. The development of national standards for food safety and quality assurance ensure acceleration of export trends as well as upgrading the quality of domestic market. Like other industries seafood export processing industry has to confront trade limiting factors of unstable economic conditions, poor infrastructure and lack of skilled workers. There is need to fill up the knowledge gap between bottom and top level of supply chain as well as value chain members on fish and fishery product safety measures.

Common problems like lack of good packaging material, bad quality cartonnes, incorrect information on the labels on the carton boxes, spots on the skin of the products due to insects and fungus are the common quality issues from Kerala. Fish boxes often are damaged due to rough handling during the transport. Lack of information regarding the packaging demands of the buyer also may be a reason for various packaging and labeling issues in the seafood export processing industry which may lead to rejection. It has also been noticed that the fish products are sometimes transported by not maintaining the proper temperature due to various reasons resulting in deteriorating the quality of the products before reaching the destination

In some instances the process used for packing materials which had not been tested for tensile strength, puncture resistance, transport worthiness etc. The poly bag used for packing tuna for export was thin and fragile. Labels used on poly bags sometimes don't maintain the demanded quality. Packing material received is to be included as Control Point in SOP and HACCP manual shall be modified accordingly. Packing materials shall be tested in an accredited laboratory for tensile strength resistance and transport worthiness. Only packing materials with suitable tensile strength, puncture resistance transport worthiness shall be used for packing tuna. Suitable training will be imparted to the purchase supervisors, who are authorized to purchase the packing materials. Laminated labels with plastic ring shall be tagged to poly bag henceforth to avoid loss of labels during transit.

US Anti-dumping Issue

India's seafood export industry is facing a serious threat after an American shrimp producer's organization has filed a petition against subsidizing shrimp export by seven countries including India, the detail of which is already been mentioned in the previous chapter. COGSI filed petitions with the US government seeking relief from the subsidized shrimp imports from China, Ecuador, India, Indonesia, Malaysia, Thailand and Vietnam. The petition seeks imposition of CVD on shrimp from these countries. Indian seafood industry faced deep crisis as the US is the largest importer of Indian seafood in value terms.

Indian Quality Standards Regulation

The Export Inspection Council (EIC) is the apex designated agency to regulate the quality standards of fish and fishery export products.EIC imposes three types of inspection and certification namely, consignment wise inspection, in-process quality control and a food safety management system based certification. EIC encourages and to facilitate worldwide access for Indian exports through a credible and efficient inspection and certification system and earn global recognition as India's premier organization for certifying quality and safety to meet international norms. All consignment of Indian Fishery products exported to the EU are required to be accompanied by a numbered original health certificate, comprising a single sheet duly completed, signed and dated. The original health certificate is required for customs clearance at the destination and shall be made available to the customs authorities at the destination before the arrival of the consignment. The consignments cannot be cleared on the basis of a copy of the original or on the basis of a fax copy of the original. Health Certificates to be given before shipment and cannot be given retrospectively. Only the officials of Export Inspection Agency are authorized to issue and sign the health certificates for exports of fishery products to EC. As per the directive it is clear that in case it is found that an establishment has obtained health certificate from any authority other than Competent Authority, approval granted to their units for exporting to EU will be withdrawn forthwith.

International food regulations are facilitating to provide clean wholesome, nutritious and safe food to the people all over the world. Harmonization of food standards is required for getting the countries of the world to agree on food codes. It is subjected to risk analysis and regional dietary classification. Seafood processing and marketing is becoming increasingly competitive globally. Aligning the quality assurance system to the international standard is indeed essential to cope with the ever demanding requirement of importing countries in the changing scenario. EIC established its network throughout India so that the requirements of the importing countries are addressed in a timely manner and the need of the exporters is fulfilled without delays so as to facilitate exports.

Seafood processing industry is one which consumes large quantities of potable water. The demand for water will be still more in view of implementation of EU requirements and HACCP programmes. Therefore, it is essential that required supply of potable water is ensured to the industry. Other inputs such as ice, insulated containers, insulated and refrigerated trucks are also needed to ensure quality of the finished products. Better infrastructure for human resource development in all fields related to fishing and fish processing should be developed. This should cover the manpower requirement for the emerging class of new design of fishing vessels and fish processing plants operating on the HACCP/EU principles.

Russia's ban on US products a boon for Indian Trade

Russia has imposed embargo on food imports from US, Canada, Australia and European Union. Indo-Russian consulting firms have come forward to help step up seafood exports to Russia. Fish export processing industry with support of MPEDA is also trying to increase marine food exports to Russia by contracting major exporters which can fetch good foreign exchange to India. According to MPEDA, India is exporting 7,400 tonnes of seafood including shrimp and other fish varieties worth about Rs. 310 crore to Russia in a year. There were 262 seafood processing establishments and 37 independent cold storages approved by the EU. Federal Service for Veterinary and Phytosanitary Surveillance (FSVPS) has so far approved 84 Indian seafood processing establishment for export of marine products to Russian Federation.

SPS, Codex Standards, US FSMA Act

SPS measures the regulations setting the maximum residue levels for toxins or contaminants, approval procedures for additives, quarantine requirements to minimize the spread of pests and diseases, labeling requirements to notify consumers of potentially-harmful foodstuffs (such as allergen-containing products), regulations governing the process or production method whereby the product is made, inspection or certification requirements or outright bans on potentially hazardous products(WTO,2005) and has already been mentioned in detail in chapter four. USFDA has brought in the Food Safety Modernization Act (FSMA) amending the existing Federal Food Drug and Cosmetics Act which in turn had amended the Bio-Terrorism Act of 2002. The same has been circulated among the exporters through the field offices. A meeting was conducted on 4th October 2012 at Chennai in which a presentation was given to the seafood exporting community on the details of FSMA Act, process of registration and other requirements laid down for the purpose of registration by the Director, Indian Office, Department of Health and Human Services, Food and Drug Administration and Embassy of the United States of America. The prospective exporters to USA have registered themselves with USA.

Health Certificate Requirement for export to Canada

The Canadian Food Inspection Agency (CFIA) brought out guidelines for import of aquatic animals on 10th December 2011 with amendments to Health of Animal Act. These new guidelines are operational from 10th December 2012. Regulation is applicable to live, chilled and frozen products of fish, crustaceans except cephalopods and products which are processed and packed for ready consumption. MPEDA offered comments to DAHDF regarding CFIA comments on the Aquatic Animal Health Certificate for export to Canada.

New Requirement for registration with Chinese Authority

The General Administrative of Quality Supervision, Inspection and Quarantine (AQSIQ) has amended Implementation Catalogue for Registration of Overseas Manufacturers of Imported Food to include seafood. All overseas enterprises desires of exporting seafood to China will now need to be registered with the certification and Accreditation Administration of China (CNCA). According to CNCA the deadline for completion of registration is 1st May, 2013. The manufacturer shall be approved by relevant competent authority in India. Manufacturer should be under the effective control and surveillance of competent authority. The sanitary conditions of the manufacturer shall meet relevant provisions of the laws and regulations, standards and codes of China.

Resolving Quality Problems in Japan

In view of rejection of Indian seafood products on account of the presence of Ethoxyquin, an ingredient to fish meal as antioxidant, a three member delegation led by Chairman, MPEDA visited Japan during 4th to 7th September 2012 to initiate dialogues with the Japanese health authorities to resolve the problem. The delegation also met Japanese Minister of Health, Labour and Welfare and requested that the MRL for Ethoxyquin in shrimp may be fixed only after appropriate risk analysis and imposition of the default level of the positive list (0.01 ppm).

Supply Side Issues

Marine products contribute a major portion to national export earnings. Customers like retailers, wholesalers, and distributors have high bargaining power because of the short life of the products. Building up strong customer partnerships, sound market research, and excellent quality of products, reliability of supply, and a constant drive for improvement, price competitiveness and attractive packaging are the key to the success of the industry. The strategic business unit has been able to bring the product to the market faster than its competitors. Shrimp have been in demand from the time when export trade practices started. Traditionally different practices existed to grow and harvest shrimps in its natural habitats in different regions. Scientific systems have been developed to culture shrimps in protected and manually controlled regimes. Development of brackish water shrimp farming and fresh water prawn farming has been well supported by the process of backward and forward integration with necessary ancillary industries.

Production wise Issues

According to joint FAO and OECD projections, world fisheries and aquaculture production is expected to reach 181 million tonnes by 2022, which is 18 percent higher than the average level for 2010-12(OECD-FAO,2013).Other than foreign exchange earnings, the fishing sector provides employment opportunities to thousands of people in the economy. Seafood processing and marketing is becoming increasingly competitive globally. Aligning the quality assurance system to the international standard is indeed essential to cope with the ever demanding requirement of importing countries in the changing scenario. The emerging global thrust in the adoption of hygienic production practices for adhering to the World Trade Organization standards; compliance to Hazard Analysis and Clinical Control Point (HACCP) by seafood exporters is a prerequisite. The supply chain in India is not strong enough to meet these rigorous standards. States such as Andhra Pradesh, Tamil Nadu, Kerala, Maharashtra, Goa, , Gujarat, West Bengal and Orissa have huge marine products potential that needs to be harnessed in a manner that can enhance India's export potential, provided all possible incentives and encouragement in terms of policies and finance is given to exporters. Good Governance of the fisheries sector demands a variety of information that leads to better management of resources.

Enhanced production and trade is unavoidable to tackle the projected gap in demand and supply of fishery products from the anticipated rise in consumption and imports due to increasing population and rising purchasing power in many countries. Marine industry in Kerala is characterized by lack of domestic competition. The rivalry among existing competitors in the international market is very high. The firms are trying to diversify production processes. The existence of a large number of firms is a feature of marine industry. Since the firms under study have been in existence for more than 40 years they are able to withstand the rivalry of competing firms. A large number of firms seem to be competing for the same customers and resources. High storage costs and perishable marine products intensify competition for customers. Supply and demand in this particular industry is so volatile that it will affect rivalry among firms. Stringent quality and environmental regulations are the major obstacles for the seafood export.

Issues in Marketing and Price Mechanism

Fish and fishery export processing industry is highly concentrated on export market and a limited percent is concentrated on domestic market with value added products. Capture fisheries account for the largest part of India's exports of fishery products. Marine products are one of the largest foreign exchange earners for India. Marine products constitute around 13 percent of India's total agricultural exports in value terms. Majority of Indian fisheries products are exported to various South East Asian markets, Japan and European Union. The export basket of fishery products is changing over the years. Marine products are facing depletion around the world and the supply is not able to keep up with the demand and the market structure of marine products is imperfectly competitive. Full capacity utilization is impossible due to non-availability of raw materials. Marketing of fish and fishery product export are highly dependent on consumer preferences which can change at any time. It is a very dynamic industry and the profitability is unpredictable.

Shrimp constitute a major share in seafood export earnings of the country. Shrimp is a short duration crop that receives high investment returns and enjoys an expanding market. The bargaining leverage and the buyer volume of marine products are very high. Buyers have access to information regarding scientific improvement in the various countries exporting marine products. The market is further characterized by changing preferences due to increasing health awareness. Prices are very sensitive. There are differences in day to day prices due to price mechanism in the open market. The success of fish and fishery export processing industry can be attained by minimizing the cost of production with high quality standards achieving economies of scale. For this marketing and distribution channels should be efficient and should keep up good industrial relations with product and process innovation. For efficient finance and administration easy access to capital effective cost control and efficient information system are needed.

The fish export processing industry is passing through a tough phase due to unforeseen developments in the international scene. A major problem has been the recessionary trends in the US and Europe. Exports to Japan hindered due to the Ethoxyquin issue and China's implementation of new regulations on exports and certification. All the shipping lines operating from Indian coasts have unilaterally announced an increase of US $ 1500 in freezer container freight rates irrespective of the size of the container and port of destination. The lack of effective legislation for fixing the freight rates of specific destinations have contributed to this unhealthy practice of shippers bargaining the freight rates. Integration of developments in advanced technologies such as information and communication technologies offer vast scope for rapid improvement and progress. Opening of global markets may lead to export of our developed technologies and facilitate generation of additional income and employment opportunities. MPEDA hires experts to build brand equity in key markets.

Processing infrastructure and value addition

During the year 2012-13 the approved Budget Estimate under the head processing infrastructure and value addition was Rs.1, 170 lakh and a total amount of Rs.1, 206.68 were spent under various components thereby recording an overall achievement of 103.14 percent. During 2011-12 an assistance of Rs. 14.44 crore was extended to 80 manufacturer exporters under the scheme for the export of value added marine products. During the financial year 2012-13 Rs. 6,57,87,403 has been disbursed to 77 manufacturer exporters. Value addition being the thrust area for increasing seafood's exports from India, it is necessary to equip the seafood processors to create state-of-art technology in handling, preprocessing, processing, packaging, warehousing and transportation. New facilities of value addition need to be created and the existing facilities needs to be expanded and additional facilities for value addition are created.

Fish is a very versatile ingredient and is suitable for all kinds of cooking, including grilling, baking, poaching, frying, in curries and kebabs. Price movement depends on demand and supply and both factors are hard to predict in the shrimp business. At present the price is fixed in the international market with the help of information technology and the negotiations for price is done through phone, email or fax. The purchase price offered by the firms to the fishermen depends to a great extent on the price that the importers are ready to offer them. The fishermen have lost their bargaining power now. Being a highly perishable commodity the farmers are not able to meet the high cost of transporting the products to better markets. And hence have no better option than to sell it to the nearest and highest bidder. Forward integration is not possible and there are absolutely no switching costs for the buyers of raw materials. Increasing scope for export of value added fishery products from developing countries to developed market is evident from the trade pattern during the past few decades. Value added fishery products marketed in consumer packs requires more demand outlets.

Marine export trends revealed that the exports which initially increased slowly due to power shortage, poor handling facilities, delays in transportation and poor communication facilities (Annon, 2005). The Indian seafood industry is annually losing Rs.6000 crore in spoilage due to poor logistics support (Annon, 2005). There is a need to find out prior identification of factors responsible for quality deterioration. The influence of different handling practices, effect of delayed icing are assessed and the impact of physical, chemical and sensory methods at various stages of processing are also investigated (Francis, 1992). High quality seafood requires the catch should be chilled immediately in ice slurry. There is need for proper landing facility and handling area in the fish landing centers. It is seen that there is a serious shortage of potable water to wash the catch to maintain the quality. Better hygiene practice should be involved in handling fish.

Technology Upgradation Scheme for Marine Products (TUSMP)

During 2012-13 an assistance of Rs. 496 lakh was disbursed to 8 processing units under TUSMP scheme for setting up new units for value addition and expansion of the existing production capacity of value added products and for diversifying into value addition by installing required machinery and equipments. Assistance for the maintenance of Cold Chain right from harvest until it is consumed. Hence necessary infrastructure needs to be created at various levels for the maintenance of Cold Chain and the quality of the fish and fish products from farm to fork. Under the head of Maintenance of Cold Chain an assistance of Rs. 220 lakh was provided to beneficiaries belonging to the various categories like the vessel operators, shrimp farmers, chilled fish exporters, preprocessing plants, processing plants etc during 2012-13.

Financial assistance of Rs. 34.65 lakh was given to beneficiaries under various categories of the schemes for proper preservation of raw materials in iced condition on board fishing vessels, in shrimp farms, peeling sheds, processing plants, and chilled fish handling centers. Financial assistance amounting to Rs. 184.5 lakh was extended to 7 beneficiaries for establishment of modern large cold storages with the help of which cold storage capacity of 9,797 Metric Tonnes having facility to store finished seafood products at -20°C was additionally created during 2012-13. The project under the FAO Common Fund for Commodities (CFC) is the first of its kind that envisages transfer technology for aquaculture, production and marketing of Value Added Products from farm raised fresh water fishes in India. The three years project has four main components, viz. market or product studies, technology transfer, investment promotion and capacity building and dissemination. Pre-shipment Credit facility provides access to finance at the manufacturing stage, enabling exporters to purchase raw materials and other inputs. Post-shipment credit enables Indian exporters to extend term credit to importers of eligible goods. Term finance is provided for setting up expansion, upgrading of production facilities and for equipment purchase for processing units.

Eco-labeling and Sustainability of Fisheries

The sustainable development of capture and culture fisheries highlighted the need for finding new methods of enhancing production. Besides Eco-labeling and Certification of fishery products by Marine Stewardship Council have proved to be few of the trade impediments. The Marine Stewardship Council (MSC) is an international non-profit organisation set up to help transform the seafood market to a sustainable basis. The MSC runs the only certification and ecolabelling programme for wild-capture fisheries. In 1996 the MSC implemented the first certification program Worldwide, more than 19,500 seafood products, which can be traced back to the certified sustainable fisheries, bear the blue MSC ecolabel.

Ashtamudi Estuary's short neck clam fishery becomes the first Marine Stewardship Council (MSC) certified fishery in India. Ashtamudi Lake, a Ramsar wetland of international importance, is the second largest estuarine system in Kerala and the clam fishery began in 1981. It supports the livelihoods of around 3000 fisher folk involved in collection, cleaning, processing and trading clams. The growth of Ashtamudi's commercial fishery was driven by demand from Vietnam, Thailand and Malaysia in the 1980s and the 1990s. By 1991, the catch peaked at 10,000 tonnes a year, but declined by 50 percent in 1993 due to overfishing. FAO Guidelines for Ecolabelling of Fish and Fishery Products from Inland Capture Fisheries was adopted on 27 May 2010 to facilitate certification and ecolabelling of products from well-managed inland capture fisheries, with focus on sustainability (FAO, 2010). Fishing industry doesn't have a coordinated conservation and fisheries policy. The states have different policies and this has an effect on the industry. With the exception of Goa, the state governments have not put in place any conservation measures.

Adoption of TQM

A large percent of the Indian seafood firms are traditionally family owned companies rather than professionally managed firms. This would result in the promulgation of old ideas of management whereby, conflict was seen as a healthy exercise which helped to build up each department's efficiency. Implementation of market orientation principles will lead to increase in business performance (Smitha, 2007).The business performance may be economic performance and non-economic performance and will help the seafood firms in gaining competitive advantage in exporting and strengthening the position of Indian seafood in the global market. Decentralization of decision making facilitates the participation of the lower level of employees and builds up their motivational levels and commitment to the firm. Thus employees are encouraged to make their own decisions so that they can deal with customers faster and more efficiently. Rewards systems help to improve and employee's morale provides encouragement and helps inculcate commitment and loyalty. Adoption of TQM will increase the market value of the product and will act as a face lift to the industry as a whole besides fulfilling its primary aim of ensuring product of good and consistent quality (Smitha, 2007).

Import of Fish and Fishery Products

The Ministry of Agriculture issued a notification on 7 July 2001, which requires animal products importers to produce a sanitary permit at the customs gate before entry into the country. License under EXIM policy is not required for the import of 125 species/groups of fish, crustaceans, molluscs and other aquatic invertebrates covered under FREE policy under the EXIM policy. Import of five groups of live fish is permitted under Restricted Policy. Import of Whale Shark (Rhincodon types) and parts and products of the species is restricted. Marine exporters are permitted to import but there are procedural bottlenecks and they have to obtain sanitary import permit which takes at least two weeks to get. In countries such as China the exporter just gives an undertaking that the imports are meant for processing and onward exports and the undertaking is accepted. The imported products are required to be tested by authorized labs here. Instead the certificate from agencies in countries from where the products are imported should be accepted. The major importing fish and fish products from India are fresh fish ,chilled fish, and feed from aquatic products mainly imported from countries such as Bangladesh, Japan and Pakistan, Norway, China, Singapore, Thailand and Republic of Korea. Fish feed is imported from countries like Thailand, Chile, Peru, Myanmar and Taiwan.

Wastage of Fish and Fishery Products

Another issue facing the industry is the waste generated while processing. For example most of the buyers want the heads of shrimps to be removed. In the case of cuttlefish only about 65 percent of the animal's body is edible. The waste thus generated is a nuisance and need to be disposed off. However, with some support from the government all the wastes can be converted into useful by- products such as bio-diesel. Anwar Hashim former president of the Seafood Exporters Association of India estimates that the industry generates about 400,000 tonnes of discards. To give a perspective of the magnitude of the waste, India exported 602,835 tonnes of fish in 2008-09.

In many countries, solid waste is recycled into fish meal plants or treated along with the municipal waste, whereas liquid waste is disposed of through the municipal sewage system or directly into a water body. Depending on the effluent polluting capacity and nature, one or several treatments can be considered. These treatments are classified into primary and secondary treatments. Primary treatments include operations designed to remove floatable and settling solids. They include screening, sedimentation, and flotation to remove oil and grease and other suspended solids. Secondary treatments comprise biological and physicochemical treatments. In biological treatments, the organic polluting matter is degraded by micro-organisms, which metabolise it into energy and biomass. Wastes from processing factories can also be converted into useful food and non-food products. Crustacean processing waste can be made into new products like chitin and chitosan. Shark cartilage, squalene etc. have affiliations in medicine and cosmetics.

Supply Chain and Value Chain Challenges

Willems et al., 2005 examined the strategies of suppliers from the buyers' perspective and the costs of intervention to assist the various developing country stakeholders to comply with international agro-food standards. Food safety forces supply chain actors to collaborate to obtain transparency in the supply chains that will guarantee a safe product for consumption. Food safety problems encountered by fish and fish products are microbiological contaminants due to lack of hygiene in the production process, residues from use of prohibited antibiotics, metal contaminants, parasites, and due to broken cold chain. Qualitatively, the costs of compliance with food safety can be summarized to unavoidable direct costs, which are tied mainly to heavy investments and higher operating costs. The more intangible hidden costs are reported to be substantial. These include costs encountered due to lack of information, innovation, and learning. Different international regulations and legislations are developed to protect the safety of consumers, to ensure fair trade practices in food trade, and to promote coordination of all food standards undertaken by international governments and nongovernmental organizations.

Efficiency in supply chains can be improved through various means, including growth in production, more efficient processing and distribution, transparent market information, proper market orientation, sustainable quality management, appropriate waste management, improved policy and resource management. Raising the efficiency of supply chains can help meet the policy and resource management. Raising the efficiency of supply chains can help meet the simultaneous challenge of reducing the costs of food to consumers, enhancing sustainability, food safety and nutrition, and increasing the revenue of supply chain participant, including smallholders. (Lem et al., 2014).

Supply Chain and Value Chains in the Fisheries Sector

The term supply chain is often used interchangeably with the term value chain. A supply chain consists of series of activities in which a product or a material is simply transferred from a starting point to an end point. Main supply chain actors in the seafood export processing industry are Fishermen or Farmers - Fishermen agents-Auctioneer-Supplier Agents- processors cum exporters. Supply chain shows the flow of fish and fishery products from their source to the ultimate user. Supply chain comprises a series of activities in which products are simply transferred from starting point to an end point. Supply chain involves purchasing, manufacturing, transportation, customer service and waste management. All the parameters in the Value chain are as same as in the supply chain, but in the value chain certain values are getting added in each stage i.e., grading, sorting, packaging or cold storage and repack the products.

Value chain adds incremental value to the product in the different nodes of a chain either by value addition or value creation. Value chains just add value to a product whereas in the supply chain, the product moves from the suppliers and dispatch to the end user. The value is then realized from higher prices and it is sold in the development of new, niche or expanded markets. The main objective of both value and supply chains is to maximize their net revenue. Supply chain and value chain issues pertains to low level of mechanization, use of unscientific method of fishing, low productivity, varying quality, safety and hygiene affected the structure of the seafood export processing industry.

Representative fisheries and aquaculture supply chain and value chains are summarized in Figure 5.1.Value Chain is the whole process involved in transferring fish from its point of production, whether by capture or culture, upto its point of consumption.

Figure 5.1: Supply Chain and Value Chain in the Fisheries Sector

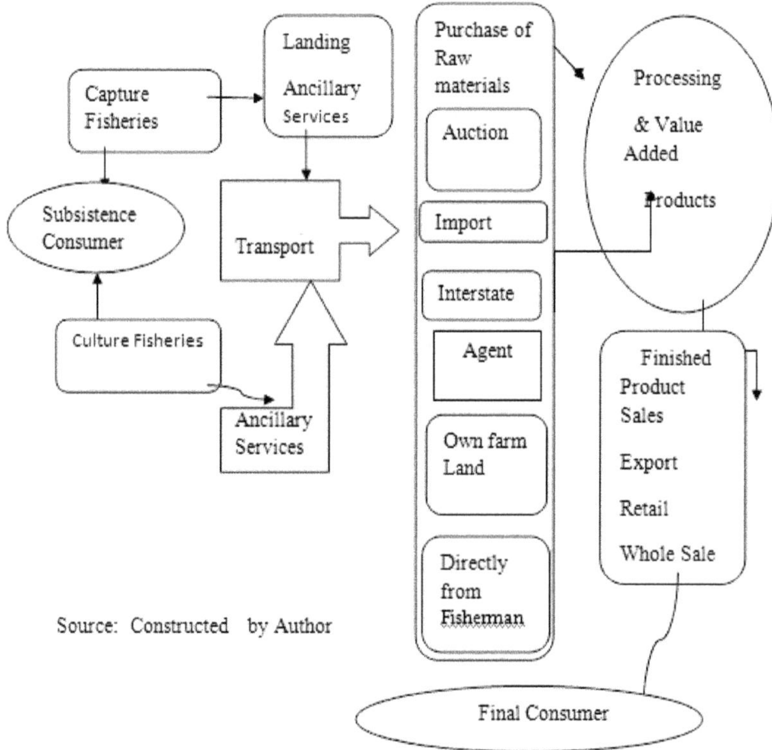

Source: Constructed by Author

Note: Ancillary Services

E.g. Boat Builders, Mechanics, fuel sellers, ice makers, gear makers, hatcheries, feed merchants, power companies etc.

There is bounding relationship between different actors and agencies found along this chain. In the domestic supply chain fish and fishery products are available either capture fisheries or culture fisheries. During 2012-13, the share of inland fish production to the total fish production of Kerala state was 22 percent (Economic Review, 2014).

A subsistence consumer purchased fish and fishery products from capture or culture fisheries. As per the population census 2011, the fisher folk population in

Kerala is 10.02 lakh covering 7.71 lakh in coastal area and 2.31 lakh in inland sector. It is also estimated that about 74100 people are engaged in fishery allied activities. The main players in the ancillary services are boat builders, mechanics, fuel sellers, ice makers, gear makers, hatcheries, feed merchants and power companies.

The main sources of raw materials for seafood export processing units are obtained directly from the fishermen or the farmer, participating in the auction, from their own farmland, supplier agents, and agents of interstate and imported from other countries. After the purchase of raw materials processors add value through processing and convert the raw materials, into value added products. Finished fish and fishery products either sold through export or distributed in the retail or whole sale domestic supermarkets and finally reached, to the ultimate consumer.

The sources of raw materials, purchased by the seafood processing export units revealed that out of the 55 survey units, 90 percent of the units participated in the auction and 85 percent of the processors have their own agent for purchasing the raw materials, 35 percent depend on interstate supply of raw materials through their agents and only 8 percent of the survey units are importing raw materials from the rest of the world. 15 percent have their own or relatives farm land and supply raw materials as joint venture and 40 percent have direct connection with the fishermen Agent and aquaculture farmers and purchase raw materials for their processing units.

The fishermen agent receives fish from the fisherman and grade each type of fish based on the freshness and quality of the fish. Auctioneer is the link between the fishermen agent and the supplier agent. Fishermen agent deals with varieties of fish whereas the supplier agent deals with particular type of products which is demanded by the fish processors. Even though our domestic fish marketing is transparent, artificial scarcity of raw material may fetch higher price which will be more beneficial to the auctioneer and supplier agent. The study observed that among the supply chain actors, living conditions of the fishermen are the lowest in the supply chain. Recently buyers i.e., importers are the price makers due to adverse effect of internet communication.

Buyers have the awareness regarding the price of raw materials and other expenditure met by the exporter and has good bargaining power in the international market with the help of their agents. Agents in between the exporter and importer are the real profit makers without any sacrificing cost. All other supply chain actors are really engaged in these fishy and frozen surroundings, their hardship, exploitation and pertinent management maximizes their net revenue. As fish is a most perishable commodity, an appropriate waste management system has to be developed to maintain hygiene and ensure proper sanitary conditions.

Understanding Fisheries Value Chain

As well as understanding the complexity of people's livelihoods, a holistic response to post-disaster assessment and recovery also requires an understanding of the entire process involved in transferring fish from its point of production, whether by capture or culture, up to its point of consumption. In the wake of a disaster, this whole "value chain" needs to be understood. Table No.5.4 shows the different types of cost for food safety and quality issues in the supply chain. Among fish processors, large differences exist with regard to costs. In particular, small-scale producers are not able to invest in technology, quality systems, laboratories, specialized staff, and training and education and depend on traders or cooperatives.

Fishing Operations

Fishing Gear, Fishing Vessels, and Infrastructure are presented under the umbrella of Fishing Operations. For the vast majority of fishing operations all around the world, a fishing vessel is needed together with its associated fishing gear. In many cases, both components are inseparable. Gear can be supplied relatively quickly to people that still possess working boats. However, for those whose boats have been destroyed, boatbuilding will take time. Building boats in a hurry will inevitably result in low-quality boats. Fishing operations require a port or landing site where the fish can be landed in hygienic conditions and then the infrastructure that allows it to be transported to market. Associated with these sites, fishers need facilities where gear can be constructed and repaired.

Table No.5.4: Cost for food safety and quality issues in the supply chain

Type of costs	Fish export processing industry
Direct costs	implementation of standards and Advanced technology.
	Auditing of Standards
	Training and education Staff
	Food Safety/quality staff
	Export Health Certificate
Indirect costs	Quality System
	Laboratory
	External Laboratory costs
	Maintenance of systems
	Visit of importers
	Relationship management
Risk	Reputation Problem
	Rejection of shipment
	Political and economic changes
	Changing quality standards

Type of costs	Fish export processing industry
	Trial and Error
	Loss of opportunities to sell products,
	Because of lack of knowledge and reliable information
Hidden Costs	Lack of innovation
	Loss during transport
	Maintenance of knowledge
	Strike of chain actors

Source: Survey data

Losses in Supply Chain

Despite the large production of food in India, food inflation and food security issues are major concerns for policy makers in the country as they affect the basic need for Indian citizens to have sufficient, healthy and affordable food, A nation-wide study on quantitative assessment of harvest and post-harvest losses for 46 agricultural produces in 106 randomly selected districts was carried out by CIPHET, Ludhiana. The different stages considered for assessment of losses are harvesting, collection, threshing, grading /sorting, winnowing /cleaning, drying, packaging, transportation, and storage depending upon the commodity. The report of the study was released in 2010.Table 5.5 depicts the percentage of losses estimated for major produces

The study has estimated that harvest and post-harvest losses of major agricultural produces at national level was of the order of Rs. 44,143 crore per annum at 2009 wholesale prices.

Table No. 5. 5: Percentage of losses estimated for major produces

Crop	Cumulative wastage (per cent)
Cereals	3.9 – 6.0 percent
Pulses	4.3-6.1 percent
Oil seeds	2.8-10.1 percent
Fruits & Vegetables	5.8-18.0 percent
Milk	0.8 percent
Fisheries (Inland)	6.9 percent
Fisheries (Marine)	2.9 percent
Meat	2.3 per cent
Poultry	3.7 per cent

Source: CIPHET Study on post-harvest losses, 2010, Ludhiana.

With adequate processing facilities, much of this waste can be reduced thus increasing remunerative wage to the producer as well as ensuring greater supply to the consumer.

Impact of Intermediaries in the Supply Chain

The marine fish market in India can be divided into two categories, domestic and export market. An efficient supply chain mechanism is essential for minimising the manipulation by intermediaries in the fish and fishery marketing channels. The scarcity of raw material reflects an increase in price of marine fish products in the supply chain system. The study observed that fishermen's share in consumer's rupee ranges from 45 to 85 percent and vary according to the type of the fishery products. It is also necessary to create proper institutional support mechanism for fishing and marketing activities as well as creation of adequate infrastructural facilities for storage and processing for improving the efficiency of fish. The main marketing channels for export oriented items like shrimps, cephalopods and fish are shown below:

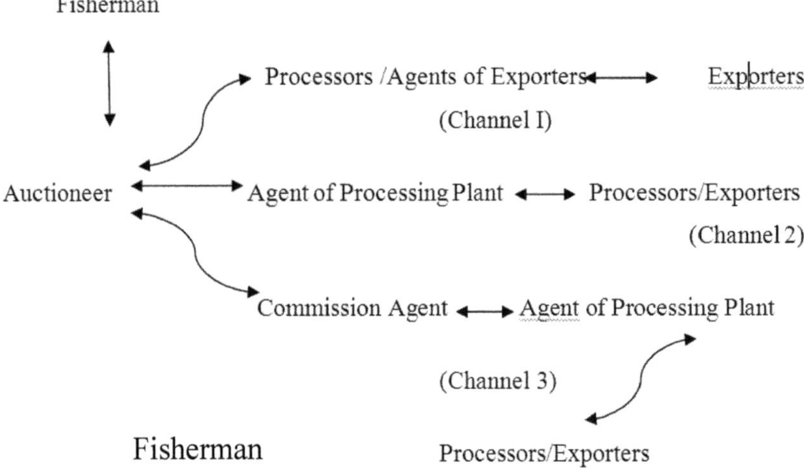

Intermediaries in the supply chain of seafood industry have a dominant role in marine fish marketing. The main supply chain intermediaries between fishermen and exporters are auctioneers, commission agents, agents of processing plant and agents of exporters.

Main Functions of Supply Chain Actors

Auctioneers: Marine fish products are sold through auctioning in majority of the coastal landing centers in Kerala. These auctioneers work on commission basis and about 1 to 6 percent of the catch value is charged as commission from the fishermen. The commission percent may vary according to the grading and nature of the fishery products. Auctioneers provide financial assistance to fishermen for their day to day operations which automatically create dependency syndrome within the fishermen. This may be one of the reasons why fishermen are not able to come out from the vicious circle of poverty. The idea pinpointed by Ragner Nurkse's balanced growth theory states that "A country is poor because it is poor" can be rightly used as "Fishermen are poor because they are poor".

Commission Agents: The commission agent purchase raw material through auctions or directly from the fishermen. They receive commission from agents of the processors, agents of exporters or wholesalers. They get around 6 percentage of the purchased value as commission from the concerned agents. They work as money lenders and also assist the fishermen in procuring diesel, ice and spare parts. They arrange for packing and transfer of the fishery product to the concerned centers. The commission agent purchases the fish at 10 to 15 percent less than the auction price.

Agent of Processors/ Exports: They receive fish from the commission agents or from auction. Their commission varies between Rs. 2 to 3 per kg. They sort fish in grades as per the quality standards requirement of the exporters/processors. They transfer fish to their processing or pre-processing units and negotiate price with the processor/exporters.

Exporter: The exporter or the manufacturer receives fish and fishery products as raw materials from the agents. Quality control technologist of the concerned processing unit test the raw material through sampling method process using the Hazard Analysis and Critical Control Point procedures and ensure the quality of the raw materials. If the test result is positive, they accept the raw material and go ahead with other processing activities. The company performs the export procedures to dispatch the raw material and negotiate the price directly with importers or indirectly through importers agent and this negotiated price with the importer may vary with season or quality and it range between 30-75 percentages of the export value. The agent of importer receives around 5 percentage of the negotiated value.

Efficiency in the Supply Chain Mechanism

The survey observed that the marine fish marketing in Kerala rely upon various marketing channels for different types of fish and fishery products. In each channel, there exist a number of intermediaries between the fishermen and the ultimate consumer, depending upon the volume of landing, sorting, storing, grading and transportation. These factors determines the cost of marketing, which finally decides the price spread or Gross Marketing Margin(GMM), Percentage Share of fishermen in Consumer Rupee(PSFCR) and Percentage share of Marketing in consumer Rupee (PMMCR).

Gross Marketing Margin (GMM) or price spread is the difference between the price received by the producer and the price paid by the consumer for any given commodity at any point of time in a market. Price spread is significant in the marine fish export marketing due to freezing cost, additional processing and value addition measures involved. Percentage share of fishermen in the consumer rupee (PSFCR) is an important indicator of the marketing efficiency. It indicates the percentage share received by the fishermen. The higher the share fisherman receives, the more efficient is the supply chain system due to lesser involvement of the intermediaries. Percentage Share of Marketing Margin in Consumers' Rupee (PMMCR) is another important indicator of the efficiency of supply chain system. It indicates the proportion of consumer's rupee that meets the cost of marketing, profit margins by the traders and other expenses involved in the marketing channel by the intermediaries. A high marketing margin indicates more number of intermediaries involved in the system and points out less efficient supply chain mechanism.

An efficient supply chain mechanism occurs when the primary producer get maximum benefit, incurring minimum marketing cost. The efficiency of the marketing can be calculated by the following equation:

1. Gross Marketing Margin(GMM)

   GMM= RP – LP

2. Percentage Share of Fishermen in the consumers price(PSFCR)

   PSFCR = LP/RP x 100

3. Percentage Share of Marketing Margin in Consumers, Rupee\(PMMCR)

PMMCR = (RP –LP)/RPx 100

Where RP is the retail price

LP is the Landing centre price

The observed data of price spread, PSFCR, PMMCR of shrimp, cephalopods and fish are shown in the Table No. 5.6

Table No.5.6: Price Spread of different Species of Marine Fish in Kerala (1Kg/ Unit)

Type	LP Rs./Kg	RP Rs./Kg	GMM	PSFCR	PMMCR
Shrimp	375	625	250	60	40
Cephalopods	130	225	95	57	42
Fish	65	175	110	37	63

Source: Survey data

Survey observed that shrimp price spread is greater than other varieties of fishery products and at the same time PSFCR share is greater in shrimp as its demand is high in the importing countries. Fish products have higher PMMCR and it shows huge margin taken by intermediaries and indicates less efficiency in the supply chain system.

Price spread varies with grades of fish products and due to the higher processing and transportation cost incurred by the intermediaries. Price spread is an important indicator of the efficiency of marketing system. Higher price spread indicates that more number of intermediaries is involved in the supply chain and it also reveals a less efficient supply chain mechanism. The marine fish pricing mechanism entirely depend upon the cartels formed by intermediaries and traders. Kerala has comparatively better supply chain system than other states. The supply of fish is highly inelastic in nature and this may lead to a price hike in future.

Profile of Seafood Export Processing Industry- Survey Analysis

This part attempts to analyse the outline of the 55 seafood export processing units on the basis the surveyed data. Analysis is conducted on the basic nature of ownership, location of units, product type, major items of exports, modern facilities, installed capacity of infrastructure and categorise workers and employers in the surveyed units. Table No.5.7 gives the ownership pattern in the surveyed units.

Table No.5.7: Nature of Ownership of Exporting Unit

Nature of Ownership	Number of Units	Percent
Proprietary	10	18.2
Partnership	21	38.2
Private Limited	21	38.2
Others	3	5.5
Total	55	100

Source: Survey Data

Location of the unit can be classified into Panchayat, Corporation and Municipality. The survey identified that 53 percent of the seafood export processing units located at Panchayat, 40 percent located at Corporation and only 7 percent located in Municipality. Out of the 55 survey units, 47 seafood export processing companies have only one production unit in the same district, 5 seafood export processing company have 2 production units in the same district, 2 units have 3 production in the same districts and 1 unit have 7 production units in the same district. Other notable reflections from the survey are that 51 seafood export processing companies have no production units in other districts. Only 4 seafood export processing companies have production units in other districts i.e. 3 seafood export processing companies have only 1 production units in other district and 1 seafood export processing company have 3 production unit in other districts. The survey identified that only 9 seafood export processing companies have production units outside the state.

Table No. 5.8 presents the type of seafood products exported from the 55 surveyed companies.

Table No. 5.8: Nature of the Product Export from the Survey Unit

Type of Product	No. of Units	Percentage
Peeled and deveined	26	47.3
Peeled and undeveined	26	47.3
Headless	24	43.6
Frozen	52	94.5
Individually Quick Frozen	25	45.4

Type of Product	No. of Units	Percentage
Value Added (Ready to eat)	4	0.073
Chilled	18	32.7
Live	5	0.09
Dried	1	0.018

Source: Survey Data

The survey results show that a very large number of units are dealing with frozen products. The Survey revealed that value-added products or ready-to-eat seafood products should be produced to meet the changing demands and consumption patterns of the consumers both domestic as well as international markets. Table No.5.9 and Figure No. 5.2 shows that major items of export items are shrimp, cephalopods and fishes.

Table No.5.9: Major items of Export

Items	No. of Units	Percent
Shrimp	38	69
Cephalopods	49	89
Fishes	48	87

Source: Survey Data

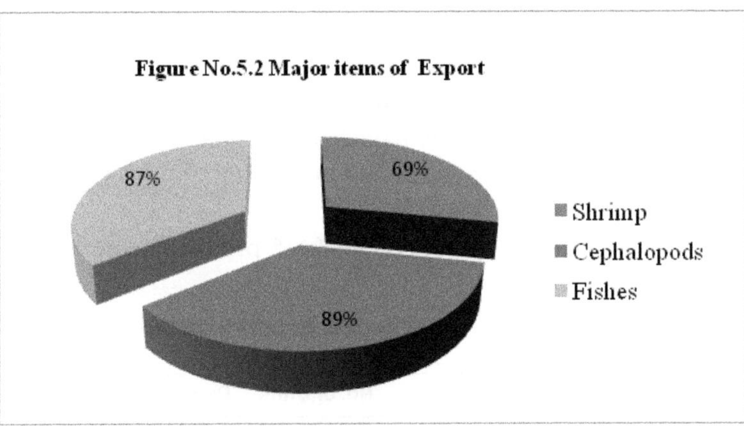

Figure No.5.2 Major items of Export

Source: Survey Data

The data revealed that seafood export processing companies are likely to export all these items according to the availability of raw materials. Basic infrastructure facilities in the export processing units should be improved so as to face international competition. Venture of modern facilities exposed in the seafood export processing industry are shown in the table No. 5.10.

Table No. 5.10: Facilities of Seafood Processing Industry

Facilities for Seafood Export	No. of Units
Refrigerated Trucks/ Containers	45
Green Technology	7
Zero Waste Approach	9
Value Addition Capacity	19

Source: Survey Data

Table No. 5.11 presents the survey results of the installed capacity of infrastructure facilities with minimum and maximum reveals the range of infrastructure capacity in these units.

Table No. 5.11: Minimum and Maximum Installed Capacity of Infrastructure

Installed Capacity of Infrastructure	Minimum	Maximum
Processing Plant MT/ Day	3	150
In- house Pre -Processing MT/ Day	4	80
Water Treatment Plant Lac Liter	0.1	50
Cold Storage Metric Tonnes	75	10000
Chill Room Metric Tonnes	2	550
Peeling Shed Metric Tonnes	3	70
No. of Insulated Fish Boxes	6	5000
Effluent Treatment Plant Lac Liter	0.1	5

Source: Survey Data

Pattern of employment in the seafood processing export industry based on the surveyed data as shown in the Table No 5.12. Out of 55 surveyed seafood units, the minimum count of employees is 15 and the maximum is 932.

Table No. 5.12: Pattern of Employment in the Firm

Categorise Employment	Total Number of Males		Total Number of Females	
	Minimum	Maximum	Minimum	Maximum
Administrative Staff	1	20	1	25
Technical Staff	1	30	1	15
Skilled Labourers	1	75	2	160
Casual Workers	2	120	2	120
Other Workers(Contract)	2	180	2	550

Source: Survey Data

An analysis of the profile of the seafood industry shows that structure of the seafood industry has changed from the traditional scenario and transformed into modernized and systematically run sector for the export market.

The survey observed that Christians followed 34 percent of the seafood export processing owners, followed by Hindus and Muslim 33 percent shared equally. The survey revealed that 73 percent of the seafood export processing companies have a brand name and 75 percent of the surveyed units have their own website space.

Problems of Survey Unit Analysis

The major problems faced by seafood export processing industry are recapitulated into seven components. These are the issues related to raw materials, labour, production, marketing, finance, administration and technical. Based on the intensity of each component, it is again subdivided and ranked using the scale from 1 to 5. These problems are encountered by the surveyed units and is analysed with the help of Kruskal Wallis H test.

Kruskal-Wallis H test

A popular nonparametric test to compare outcomes among more than two independent groups is the Kruskal Wallis test. The Kruskal Wallis test is used to compare medians among k comparison groups ($k > 2$) and is sometimes described as an ANOVA with the data replaced by their ranks. The null and research hypotheses for the Kruskal Wallis nonparametric test are stated as follows:

H0: The k population medians are equal versus

H1: The k population medians are not all equal

The procedure for the test involves pooling the observations from the k samples into one combined sample, keeping track of which sample each observation comes from, and then ranking lowest to highest from 1 to N, where $N = n1+n2 + ...+ nk$.

Test Statistic for the Kruskal Wallis Test

The test statistic for the Kruskal Wallis test is denoted H and is defined as follows:

$$H = \left(\frac{12}{N(N+1)} \sum_{j=1}^{k} \frac{R_j^{\,2}}{n_j} \right) - 3(N+1)$$

where k=the number of comparison groups,
N= the total sample size,
nj is the sample size in the jth group
and Rj is the sum of the ranks in the jth group.

If the observed value of H is greater than or equal to the critical value, we reject H0 in favor of H1; if the observed value of H is less than the critical value we do not reject H0. To determine the appropriate critical value we need sample sizes (n1=3, n2=5 and n3=4) and our level of significance is (□=0.05). We reject the null hypothesis in favor of the alternative hypothesis if any two of the medians are not from this test. The study found identifies the core issues with ranking and ensure and identified the main problem in the seafood industry and helps to resolve it amicably for its smooth functioning and lead to better economic performance.

Raw Material Problems

The first and foremost problem faced by the industry is the scarcity of raw material, its inferior quality, delayed supply and unexpected price variations.

Table No.5.13: Raw Materials Issues: Kruskal Wallis H Statistic Analysis

Raw Material	Sum of Ranks	Mean Ranks
Scarcity of Raw Materials	7017.5	127.5909091
Inferior Quality	2863.5	56.14705882
Delayed Supply of Raw materials	3707.5	72.69607843
Price Increase	8989.5	163.4454545
Test Result		
H Test Statistic	103.7477895	
Critical Value	7.814727764	
P-Value	2.42946E-22	
Reject the Null Hypothesis		

Source: computed from surveyed data.

Test results shows that H test Static is greater than the critical value so we reject the Null Hypothesis. Among the issues, price increase is the main issue related to raw materials; secondly scarcity of raw materials, then delayed supply of raw materials and finally inferior quality of raw materials. The issues are ranked according to the survey data analysis. As mentioned, the most difficult situation faced by the industry is the fluctuations in price may be due to price mechanism- change in the demand for and supply of fishery products, dollar value fluctuations, changes in the political, social and economic structure and its changes due to share market price variations. This price fluctuation may be one of the sources of exploitation among the supply chain actors and may lead to an artificial scarcity of raw materials.

Labour Issues

The second component is labour issues are sub-divided into shortage of. trained labours, lack of professional labour and trade union problems

Table No.5.14: Labour Issues

Labour	Sum of Ranks	Mean Ranks
Shortage of Trained Labours	5212.5	94.7727273
Lack of Professional Labours	4401.5	86.3039216
Trade Union Problems	1412	33.6190476
Test Result		
H Test Statistic	54.36386	
Critical Value	5.991465	
P-Value	1.57E-12	
Reject the Null Hypothesis		

Source: computed from surveyed data

Test Result shows, that H test Static is greater than the critical value so we reject the Null Hypothesis. Shortage of Trained Labours is the core concern faced by seafood industry. Availability of local labours is significantly limited as Kerala's occupational structure changes due to higher education status and on the other hand opportunity cost is higher in other industries like textile, construction, other manufacturing industry. Hence seafood export processing industry depends on interstate labours from Bihar, West Bengal, Orissa, and Assam. These labours are unskilled and are initially trained by the seafood industry but later move to other employment sectors due to higher wages, and it creates additional financial burden for training and instructing the new labourers again. The second rank is the issue related to lack of professional labours and third ranking issue is the problems of trade unions which is not very significant as majority of the labourers are from other states.

Production and Technical Issues

Issues on production are classified into higher demand, lack of energy conservation measures, lack of new technology and lack of value added products

Table No. 5.15: Production Problems

Production	Sum of Ranks	Mean Ranks
Unable to meet Demand	5921.5	111.7264151
Lack of Energy Conservation measures	3887.5	82.71276596
Lack of New Technology	4936.5	96.79411765
Lack of Value Added Products	5555.5	111.11
Test Result		
H Test Statistic	8.224800179	
Critical Value	7.814726664	
P-Value	0.041587197	
Reject the Null Hypothesis		

Source: computed from surveyed data

Test Result shows that H test Static is greater than the critical value and so we reject the Null Hypothesis. The major problem found in the production sector is the fluctuation in the demand, lack of raw materials according to the demand of the buyer's perspective. Second issue is related to lack of value added products like ready-to-eat and cook products. The modern customer demand better quality and ready- to- cook products. Third ranking problem is related to the new technological adaptation that reduces the wastage of energy and time. The energy conservation measures are demanded by future generation in order to generate a sustainable development in the fishery sector. Firm should adopt the solar energy system, power saving technology, innovate cost and energy saving methods of production to achieve maximum profit margin.

Marketing Problems

The fourth and most important component is 'Marketing issues' and is sub -divided into competition from branded firms, lack of export orders, lack of market infrastructure, changes in the market trends, delay in the approval of importing countries, and lack of quality control measures.

Table No. 5.16 Marketing: Kruskal H Statistic Analysis

Marketing	Sum of Ranks	Mean Ranks
Competition from branded firms	11120	205.925926
Lack of export orders	7001	132.09434
Lack of market infrastructure	8007	148.277778
Changes in the market trends	11913	216.6
Delay -Approval of importing Countries	8241	152.611111
Lack of quality control measures	6368	117.925926

Test Result	
H Test Statistic	49.60686
Critical Value	11.0705
P-Value	1.67E-09
Reject the Null Hypothesis	

Source: computed from surveyed data

Test Result shows that H test Static is greater than the critical value, so we reject the Null Hypothesis. Holistic marketing concept can be applied for expanding fish and fishery product export. The core issue in Marketing is the changes in the market trends and the competition from the branded firms. These issues are mainly confronted by the seafood processing export units. Hence, the exporter should be aware of the risk elements, marketing returns, current trends and opportunities in the marine export. Third issue in this area is lack of market infrastructure and fourth issue is the delay in approval from the importing country. To avoid such problems, well a structured market infrastructure is must and proper documentation and relevant clearance certificates should be prepared according to the requirement of the importing countries. Lack of export orders is another issue faced by the industry and this is interlinked with lack of quality control measures. Once the company's reputation gets strained due to quality issues of rejection and detention they automatically face the related issues in marketing. To avoid such circumstances total quality management is a necessary condition in the seafood export processing industry.

Financial Problems in the Seafood Industry.

Finance is the backbone of the seafood industry. The main allied issues related to it are taken as lack of funds for expansion, high interest rates, delay in official procedures, and lack of returns from investment and insurance claims.

Table No. 5.17: Finance Problems: Kruskal H statistic Analysis

Finance	Sum of Ranks	Mean Ranks
Lack of funds for expansion	7448	137.925926
High Interest Rates	9538.5	173.427273
Delay in Official Procedures	8132	147.854545
Lack of returns from Investment	7314.5	132.990909
Insurance claims	4695	88.5849057
Test Result		
H Test Statistic	33.05715	
Critical Value	9.487729	
P-Value	1.16E-06	
Reject the Null Hypothesis		

Source: Computed from surveyed data

Test Result shows that H test Static is greater than the critical value so we reject the Null Hypothesis. The main issues in the finance sector are the high interest rates and the delayed official procedures which definitely affect rolling of cash trend in the seafood business. Third ranking problem is the lack of funds for expansion, especially to convert the traditional frozen pattern to value added product like ready to pan implementation requires huge investment and it will return its profit margin in the long run only. That leads to lack of return from investment in the short run and resulted in a financial dilemma. Insurance claims are not a significant problem if the official documentation procedures are perfect otherwise the exporter has to face losses in the way of insurance claims.

Administrative and Technical Problems

The important issues of administrative and technical issues are taken as shortage of operational efficiency, changes in the international quality standards, high administrative cost, lack of trained technical personals, low research and development process, lack of effective communication skills and lack of waste management practices.

Table No.5.18: Administrative and Technical Issues: Kruskal H statistic Analysis

Administrative and Technical	Sum of Ranks	Mean Ranks
Shortage of operational efficiency	9667	179.018519
Changes in the international quality standards	13484.5	245.172727
High administrative cost	11150.5	202.736364
Lack of trained technical personals	9787	184.660377
Low research and development process,	10350.5	195.292453
Lack of effective communication skills	9896	183.259259
Lack of waste Management	7674.5	139.536364
Test Result		
H Test Statistic	27.36008	
Critical Value	12.59159	
P-Value	0.000124	
Reject the Null Hypothesis		

Source: Computed from surveyed data

In the administrative and technical side, the main problems are the changes in the international quality standards. The seafood industry should always be updated on the changing standards and should be able to prepare relevant certification and documentation accordingly. This leads to additional appointments of staff for documentation and other dealings with international agents and thus increases the administrative cost expenditures. Another problem is the lack of research and development for the analysis of their development. As the marine export industry deals with cultural and language diversification, it should ensure that language proficiency exist for it employees, or else it creates various related issues and countenance of additional risk to the industry. Another problem is the shortage of operational efficiency and that can be overcome through proper training and resource management both these are interlinked.

Quality Issues

The last but the most relevant component is quality issues linked with HACCP training, water scarcity, power scarcity and technological upgradation.

Table No.5.19: Quality Issues: Kruskal H statistic Analysis

Quality Issues	Sum of Ranks	Mean Ranks
HACCP Training	3315	97.5
Water Scarcity	1397.5	42.3484848
Power scarcity	4198.5	63.6136364
Technological up gradation	4198.5	63.6136364
Test Result		
H Test Statistic	-323.651	
Critical Value	7.814728	
P-Value	0.031728121	
Reject the Null Hypothesis		

Source: computed from surveyed data

The foremost ranking issue identified power scarcity and technological upgradation and both are interrelated and as for the technological upgradation power supply is most requisite factor. The next issue is HACCP training which is mandatory for the functioning of marine export industry. Training labourers is another issue. Language barriers are a crucial problem. Workers are from other state. Other issue is water scarcity which would be a key problem in future as it is observed that there is a shortage of potable water in all stages of processing from catch to table.

The Table No 5.20 scrutinizes the problems faced by in the surveyed units and ranked with the help of Kruskal H statistic analysis.

Table No. 5.20: Problems Faced in the Surveyed Unit: Kruskal H statistic Analysis

Rank	Problems of Firms	Sum of Ranks	Mean Ranks
1	Quality Issues	18101	329.11
2	Raw Materials	15314.5	278.45
3	Finance	13813	251.15
4	Labour	12532	227.85
5	Administrative and Technical	11317.5	205.77
6	Production	10985	199.73
7	Marketing	10740.5	195.28
Test Result			
H Test Statistic		129.660199	
Critical Value		14.06714045	
P-Value		7.40E-25	
Reject the Null Hypothesis			

Source: Computed from surveyed data

Out of the seven parameters, quality Issues is identified as the major problem faced by the seafood processing industry. The second key problem recognized by the analysis is the raw material scarcity and the third problem is finance. The fourth problem is identified as Labour and fifth as Administrative and Technical issues. The sixth and seventh rank identified as production and the least rank examined as issues related to Marketing.

Capacity Utilization

Capacity utilization and cost of production are the fundamental issue of the fish export processing industry in Kerala. Capacity utilization has a direct, bearing on resource generation for economic development (Shrivastava, 1992). Low Capacity Utilization results in low economic performance of fish processing export industry in Kerala. Table No.5.21 shows the number of seafood processing plants with their capacity between 2005 to 2013 from India and Kerala.

Capacity utilization was worked out to be only 46.47 percent. The main reasons for low capacity utilization were reported to be due to non-availability of raw materials, high cost of production, shortage of power, scarcity of ice and potable water.

Table No.5.21: Region-wise Processing Plants With Capacity

Year	Kerala		India	
	No	Capacity	No	Capacity
2005	81	2370.83	261	10259.93
2006	85	2542.73	278	11096.33
2007	90	2770.73	295	12004.53
2008	95	2857.63	328	13540.93
2009	97	2882.63	349	14449.69
2010	102	2976.91	374	15451.15
2011	75	3021.31	391	16503.8
2012	111	3279.81	423	17477.04
2013	112	3309.91	447	18520.48

Source: Software Design & Hosting credits: NIC Lakshadweep

The surveyed units selected are classified on the basis of their installed capacity which is given in tonnes per day and is presented in Table No.5.22. Figure 5.3 shows the percentage of capacity utilization of the surveyed units. It is observed that the range of capacity utilization is between 20 to 90. The average capacity is 46.47.

Table No. 5.22: Installed Capacity of the Surveyed Export Processing Units

Installed Capacity	Number of Firms	Percent of Firms
Less than 25	15	27.3
25-50	25	45.4
50 and above	15	27.3
Total	55	100

Source: Survey Data

Labour shortage was also reported to be a reason for low productivity during peak season. Studies conducted by Ramachandhran (1988) also showed low capacity utilization and productivity in the seafood factories in Kerala. Analysis for the capacity utilization across different quarters showed that during the period from October - December months it was 30.39 percent followed by January- March at 28.29 per cent. The processing plants processed minimal quantities during July-August and April-June. The average quantum of marine fish products processed per processing plant was found to be 2,781.70 tonnes per annum (Shyam, 2012). The major causes for the idle capacity of the surveyed units observed as the scarcity of raw materials with rise in price variation, high cost of production, inadequate power supply, shortage of labourers, fluctuating trend of the foreign markets, competition among the exporting firms for procuring the raw materials, lack of proper infrastructure facilities', investment spend in holding the material upto shipment and delayed purchase orders which leads to underutilization of plants. The best way to solve this scenario is by increasing the availability of raw material through sustainable fish production, product diversification, and adaptation of new processing techniques and through proper packing for the export to compete the challenges in the international trade.

Figure: 5.3 Capacity Utilisation Percentages

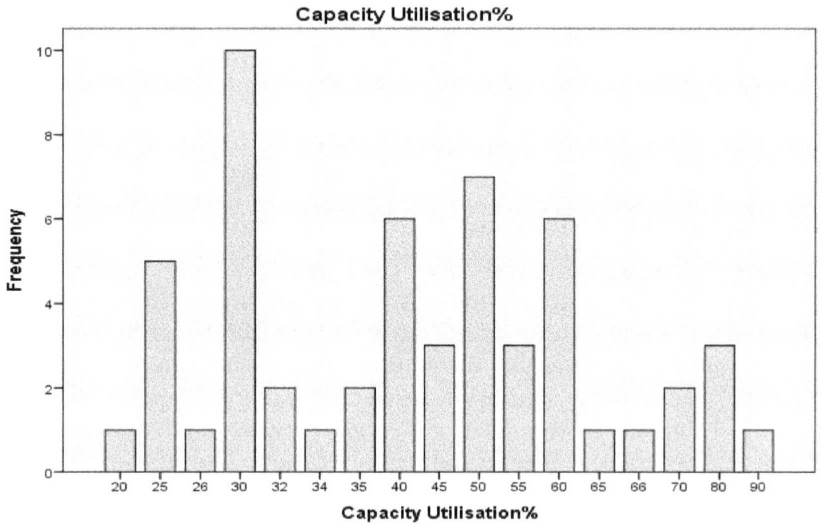

Source: Survey data

Table No.5.23 shows the export processing plant capacity utilization from the surveyed units grouped on the basis of installed capacity. The result of ANOVA proved that there are significant differences in the mean total capacity utilization of seafood export processing industry based on their installed capacity.

Table No 5.23: Processing Plant Capacity Utilisation: ANOVA

		Sum of Squares	df	Mean Square	F	Sig.
Capacity Utilisation	Between Groups	2025.602	2	1012.801	3.864	.027
	Within Groups	13631.307	52	262.141		
	Total	15656.909	54			

Source: Computed from the Surveyed Data

Table: 5.24: Multiple Comparisons Post Hoc Test Least Significance Difference Test

Depe Indent Variable			Mean Difference (I-J)	Std. Error Lower Bound	Sig. Upper Bound	95% Confidence Interval	
Capacity Utilisation	below25	between 25 to 50	-11.853*	5.288	.029	-22.46	-1.24
		above 50	-15.467*	5.912	.012	-27.33	-3.60
	between 25 to 50	below25	11.853*	5.288	.029	1.24	22.46
		above 50	-3.613	5.288	.497	-14.22	7.00
	above 50	below25	15.467*	5.912	.012	3.60	27.33
		between 25 to 50	3.613	5.288	.497	-7.00	14.22

* The mean difference is significant at the 0.05 level.

Source: Computed from the Surveyed Data

LSD-Post Hoc Test highlighted 6 possible pair-wise comparisons shown in the Table No. 4.24 If the significant value is greater than 0.05, for units between 25 and 50, above 50 and it indicates that those firms are not statistically different in their capacity utilization. If the significant value is less than or equal to 0.05 it shows that there is a statistically significant difference between the two conditions being compared and is shown in the mean difference column highlighted by the asterisk sign. The survey group below 25 (between 25 and 50, above 50) shows statistically significant difference in the capacity utilization from the surveyed industry. The results highlighted that capacity utilization is higher in medium and high installed survey units than low installed capacity units.

Export Documentation Procedures

The exporters have to obtain PAN based Business Identification Number (BIN) from the Directorate General of Foreign Trade prior to filing of shipping bill for clearance of export goods. Shipping Bill/ Bill of Export is the main document required by the Customs Authority for allowing shipment. Along with the Shipping Bill, other documents such as copy of packing list, invoices, export contract, letter of credit, etc. are also to be submitted. There are 5 types of shipping bills.

1. Shipping Bill for export of duty free goods. This shipping bill is white colored.
2. Shipping bill for export of goods under claim for duty drawback. This shipping bill is green colored.
3. Shipping bill for export of duty free goods ex-bond i.e. from bonded warehouse. This shipping bill is pink colored.
4. Shipping Bill for export of dutiable goods. This shipping bill is yellow colored.

5. Shipping bill for export under DEPB scheme. This shipping bill is blue in colour.

Export documents can be classified into two. They are commercial documents and regulatory documents. Out of the fifteen commercial documents, eight are principal and the rest are auxiliary. The eight principal documents are the commercial invoice, packing list, and bill of lading/air way bill, certificate of inspection/quality control, certificate of origin, bill of exchange, shipment advice and insurance certificate. The insurance document evidencing insurance has been taken out on the goods shipped, and it gives full details of the insurance coverage. An insurance certificate certifies that the shipment has been insured under a given open policy and is to cover loss of or damage to the cargo while in transit. This insurance will not cover the quality issues of seafood products. The seven auxiliary documents are proforma invoice, intimation for inspection, shipping instructions, insurance declaration, and application for certification of origin, mate's receipt, and letter to bank for collection/negotiation of documents. There are seven Regulatory documents associated with the pre-shipment stage of the export transaction. They are ARE Form (for Central Excise), Shipping Bill/Bill of Export (for Customs), Port Trust Copy of Shipping Bill/ Export Application/Dock Challan. Vehicle ticket, Exchange Control Declaration/ GR/PP forms, Freight Payment Certificate, Insurance Premium Payment Certificate. After shipment, documents are to be submitted to bank for realizing export proceeds.

Reasons for the Delay in Pre-shipment Clearance

The importer is generally responsible for arranging the pre-shipment inspection; the exporter must make the goods available for inspection in the country of origin. Delays in the process can lead to problems with the shipment and increased costs for the exporter. Therefore, it is in the best interest of exporters to work with their freight forwarder to ensure that all information is accurate and is provided to the inspection company immediately after notification of the requested inspection. If a disagreement arises on the findings of the pre-shipment inspection, a resolution to the discrepancy should be negotiated with the inspection company. However, if exporting to a World Trade Organization (WTO) member country, the WTO Agreement on Pre-shipment Inspection spells out the responsibilities of the exporter and the inspection company. The agreement requires the inspection company to appoint an appeals official and comply with the agreement guidelines when carrying out their pre-shipment inspection services for signatory countries. Sanitary and phytosanitary certificates, which are normally issued to protect U.S. consumers, can also be used for international trade purposes. The Certificate of Origin is issued by the competent authority in the exporting country.

Survey examines the reasons for the delay of Pre-shipment Clearance using the scale from 1 to 5 and analysed it by using the kruskal H Statistics method. Table No.5.25 identified the major problems faced by the Pre-shipment clearance from the surveyed unit. There are 9 parameters chosen for problems related to pre-shipment clearance and are listed with ranks.

Table No 5.25: Problems Faced in the Pre-shipment Clearance: Kruskal H statistic Analysis

Reasons for the Delay in a Pre-Shipment Clearance	Ranks	Sum of Ranks	Mean Ranks
Random Opening of Self-Sealed Containers	4	13532	255.3
Non-Availability of Custom Officials for Inspection	8	12587	228.9
Congestion at Port	2	16676.5	308.8
Poor Port Handling Facility	3	14862	280.4
Lack of Adequate Berthing Space in Ports	6	12835	242.2
Missing Container Trailers	9	3153.5	59.5
Processing of Shipping Bills	7	12639.5	229.8
Storage and Warehouse Facilities	5	13427	253.3
Corruption Practices	1	17173.5	318
Test Result			
H Test Statistic			124.768254
Critical Value			15.5073131
P-Value			3.43E-23
Reject the Null Hypothesis			

Source: Computed from the survey data

Survey examines the reasons for the delay of Pre-shipment Clearance using the scale from 1 to 5 and analysed it by using the kruskal H Statistics method. Table No. 5.25 identified the major problems faced by the Pre-shipment clearance from the surveyed unit. There are 9 parameters chosen for problems related to

The foremost problem identified is corrupt practices which are all pervading in commercial procedures and documentation and other coordination activities of pre-shipment clearances. The study suggested that all document should be minimized and shipping permitted on invoice basis which involves all requisite information. In order to overcome, the problem of corruption practices, government should reduce discretionary powers of the concerned officials. Second and third ranked problems are related to the poor infrastructure facilities. Random opening of self-sealed containers is ranked as fourth. Other problems are storage and warehouse facilities, Lack of adequate berthing space in ports, processing of shipping bills, and non-availability of custom officials for inspection and missing container trailers. Electronic processing of document procedures should be made more effective by better training of officials and ECGC should be more active.

Management Issues Related to Employees

Out of the 55 surveyed units, 39 units (71 percent) revealed that they are facing employee's related issues. These issues are classified into eleven categories and ranked using the scale from 1to 5. All the parameters are analysed by using the Kruskal-Wallis H test. Table No.5.26 examines labour issues according to the ranking scores.

Table No.5.26: Management Issues Related to Employees: Kruskal H statistic Analysis

Labour Issues	Rank	Sum of Ranks	Mean Ranks	Sample Size
Wage Rate	3	15413.5	280.2454545	55
Social Security Scheme	2	14890	280.9433962	53
Unexpected Contingencies	1	14277	291.3673469	49
Labour Strikes	11	2321	116.05	20
Industrial Relation	4	14115.5	261.3981481	54
Work Environment	9	12171	234.0576923	52
Canteen Facilities	10	8178.5	221.0405405	37
Job Security	5	13060	256.0784314	51
Gender Discrimination	8	10776.5	239.4777778	45
Working Hours	6	13266.5	250.3113208	53
Conveyance	7	8790.5	251.1571429	35
Test Result				
H Test Statistic	28.2317			
Critical Value	18.307			
P-Value	0.00166			
	Reject the Null Hypothesis			

Source: Computed from the Survey data

Test results shows that H test Static is greater than the critical value so we reject the Null Hypothesis. The main issues faced by the seafood export processing units are ranked, first for unexpected contingencies, second due to lack of social security schemes and the wages ranked as third. All these issues are interlinked and exporter revealed that moderate range is applicable to solve these most relevant issues without which industry cannot run competently. Fourth rank is rated for the industrial relations, i.e. the lowest grade labours to the highest grade in management should have a healthy communication regarding wage revision, bonus, and service conditions of the workers and having a minimum wage fixation and can ensure a positive industrial status of the unit. The Industrial relation aspect observed that lack of healthy relationship between the bottom levels to top level adversely affects the overall progress of business in the surveyed units. Other factors based on their ranks are listed as job security, working hours, conveyance, gender discrimination, work environment, canteen facilities and labour strikes. It is observed that highest turnover seafood companies provide better facilities than the lowest turnover companies. Usually labour strikes are not common in the seafood export industry as the unit deals with most perishable commodity. Most of the labourers are from outside the state and are in tied up with agents on a contract basis and have no direct relation with the owner of the surveyed unit.

Economic Recession

Economic recession caused the liquidity crunch in the world economy. Shortage of liquidity in various countries adversely affected the aggregate demand of various goods and services. In order to face the problem of availability of liquidity, US and EU countries had to withdraw their investments in developing countries. The worldwide recession has also taken a toll on the fishing industry of Veraval in Gujarat. Veraval, which is considered to be the hub of fishing exports, is now going through its worst possible phase, as fish traders report a dip of almost 30 percent in overseas demand. Fisheries are one of the main export oriented Industries in India, as it earns billions of dollars from world markets through its fish and other seafood exports (Singh, Yadav, 2009). The encouraging results from the advance estimates for 2014-15 suggest that though the global sluggishness has partly fed into the lackluster growth in foreign trade; this downward pressure has been compensated by strong domestic demand, keeping the growth momentum going.

Out of the 55 surveyed units, 39 companies revealed that global recession partially affected seafood export processing industry. There are ten factors which are taken in 1 to 3 point scales, graded the global recession and analyzed it with the help of Kruskal H Statistics. Table No. 5.27 presents the affected factors of global recession in the surveyed units with their ranking grade. All the factors are not equally affected and the effect of variation can be exposed with the help of sum of Ranks or their mean ranks. The highest Mean rank factors are the fluctuating foreign exchange rates and prices which were affected more in the global recession. Changes in the dollar rates partially affected the prices of seafood products during the global recession. Third ranking grade factor is the demand for marine fish products, for example, the European Union countries Italy and Germany demanded more cheap rate fishery products than high rated fishery products as revealed from the surveyed units. The fourth ranking factor i.e., Client Market and their Payment Schedules delayed due to global recession. Global recession partially changes the export destination and was ranked fifth. Sixth ranked factor is the changes in value addition and the seventh ranked factor is modification of factory operational plans which also partially got affected due to global recession. Least affected factor is the value chain management, which revealed that at each stage of production, additional value is added to withstand the global recession and did not adversely affect the surveyed units.

Table No. 5.27: Factors Affected Global Recession in the Marine Export Industry Kruskal H statistic Analysis

Factors Affected Global Recession	Rank	Sum of Ranks	Mean Ranks
Production Capacity	9	11218.5	203.97
Price	2	19364	352.07
Client Market and their Payment Schedules	4	17133	311.51
Export Destinations	5	16254	295.53
Fluctuating Foreign Exchange Rates	1	20615.5	374.83
Demand for Marine Fish Products	3	18688	339.78
Changes in Value addition	6	14969	272.16
Effect on Employment	8	11455	208.27
Modification of factory operational plans	7	11488.5	208.88
Value Chain Management	10	10339.5	187.99
Test Result			
H Test Statistic	94.30518948		
Critical Value	16.91897762		
P-Value	2.21947E-16		
Reject the Null Hypothesis			

Source: Computed from the surveyed units

Problems and Challenges Faced by Survey Unit.

Improving Quality Standards by implementation of National and International Standards, promote brand in the Market, improve sanitary conditions, minimise losses and change the working atmosphere of the unit are the foremost challenges faced by the seafood export processing industry. Analysis of the survey observed that the third party inspections on quality, facility, social ethical are costly. Survey analysis showed that social ethical sourcing audits are to be included in the firm. Sanitary import permit become mandatory. It is the current requirement by importing countries due to new regulations. Industry need to follow and avail the sanitary permit. 100 percent of the survey units have an assimilation to keep themselves updated to all quality standards information for a stable running. Out of the 55 surveyed units 50 units i.e., 91 percent revealed that seafood firms incurred additional cost due to inspection. 69 percent agreed that the seafood export processing industry faced sudden changes in rules, laws, regulations and policies.

Among the surveyed unit 75 percent are satisfied with the measures taken by the government regulatory authorities to face the challenges in export industry. 84 percent of the survey units have well equipped lab and testing facilities to ensure proper quality drills and material traceability. Only one survey unit have less than or equal to 50 percent ownership and also one survey unit revealed that have more than 50 percent foreign ownership. Out of the 55 survey units, only 4 units import raw materials from other countries. 26 survey units import technology from the rest of the world. 39 survey units pointed out that the seafood industry faced labour issues. Survey revealed that 95 percent of the survey unit's wastes are taken by Private Agents, 2 percent of the unit waste are taken by Government. 3 percent of the waste is used for processing and marketing of marine by products.

Analysis of the survey revealed that the main problem faced by seafood export industry is scarcity of raw materials and its uncertain price rise. The core problem faced by processing industry is shortage of trained manpower, waste water management, power failure, ice shortage and lack of commitment of outside workers. Labour shortage is severe because skilled labours are located in Kerala than the northeastern states. The industry has to face stringent import norms by EU and USA and competition from other ASEAN seafood processing countries. Lack of trained manpower especially non-availability of workers for value added production.

Survey revealed that financial, technical and training assistance from state and central government are moderate. Contributions from fisheries Universities are low. CIFT provides training for HACCP and other quality issues faced by the industry. Assistance from EIA and MPEDA are contributed significantly for the welfare of seafood export industry. Survey remarked that low assistance from private and international institutions. All Surveyed units are availing duty pay back scheme and Vishesh Krishi Gram Udyog Yojana scheme. Out of the 55 survey units, 69 percent of the survey units revealed that the seafood firm faced delay in the export incentives disbursed by DGFT.

Government Legislation

The Government of Kerala gives top priority to the fisheries sector, because this sector provides employment and income to more than one million people, either directly or indirectly. Fishing industry satisfies the protein requirements of a considerable chunk of the population. Fisheries sector provides considerable revenue, especially in foreign exchange, to the exchequer of the State. Hence, it has undertaken several projects and programmes for increasing production, for conserving and ensuring sustainable exploitation of fisheries wealth, for promoting cultivation of fish and prawns, for development of fishing harbours and facilities for landing of fish, for strengthening facilities for marketing of fish, and for the upliftment and welfare of the fisher-folk. The importance of fisheries sector in Kerala and the advantageous position that Kerala enjoys as a maritime State is to be considered in the national policies. There is a great need to introduce recent trends in fish processing technology to maintain the state's position in international market. India was emerging as a major marine product exporter. The fishing industry not only acts as foreign exchange earner but also plays an important role in Kerala economy. But with the development of technology and liberalization of international trade, the foreign countries adopted trade barriers to limit Kerala's fishery product export, which had made Kerala fishery product export disproportionate with fishery products production.

The government procurement process is not competitive and often lacks transparency. Information about government projects and procurements is often not readily available from the appropriate authorities and interested parties must spend considerable time to obtain the necessary information (Government has undertaken several policy initiatives and measures to boost the growth of fisheries industry in India. At the Central level an important policy has been announced as the Comprehensive Marine Fishing Policy, 2004. The objective of the policy is to augment marine fish production of the country up to the sustainable level in a responsible manner so as to boost the export of sea food from the country as well as to increase the per capita fish protein intake of the masses. The policy, thus, advocates protection, consideration and encouragement of subsistence level fishermen. It seeks to promote conservation, management and sustainable utilization of India's invaluable marine wealth. The Union Government also reduced the Central grant component for infrastructure development from 70 percent to 40 percent. Assistance from National Fisheries Development Board has been reduced to 40 percent from the previous level of 90 percent. The State Coastal Development Corporation is implementing projects worth Rs. 580 crore for the development of the coastal areas in the State.

Other Issues

Seafood exporters from Kochi port are facing innumerable difficulties, following the upgradation of the Customs EDI software from version 1 to 1.5. The technical snag that originated in August 2010 has stalled the disbursement of export incentives to exporters not only to Kochi, but across the country. The Kochi port is one of the worst affected due to the system upgradation. The other major ports such as Chennai, JNPT and Mumbai have not been adversely affected as the Customs EDI version has not been converted to the new system. The officials in the Seafood Exporters Association of India (SEAI) pointed out that export benefits worth Rs 100 crore have been blocked on account of this .The shipping bills filed by the exporters/Custom House Agents and Export General Manifest (EGM) filed by the shipping lines with the Customs Electronic Data Interchange (EDI), have to be uploaded in the DGFT site automatically, and the export incentives are processed by DGFT based on these transferred data. The export incentives granted under schemes such as Duty Entitlement Pass Book scheme as converted as Duty Drawback Scheme, Vishesh Krishi Gram UdyogYojana and Export Promotion Capital Goods scheme, are not fully functional. Over 3,000 such shipping bills are pending with DGFT offices all over the country. Inspite of the various representations by the Exim trade bodies to the higher authorities at the Ministry of Finance, Commerce and to the System Directorate of Customs at New Delhi, no effective steps have been taken to rectify the problems.

CONCLUSION

Fish processing and marketing are becoming increasingly competitive globally. International quality assurance standards and regulations are indeed important to cope with the ever demanding requirement of importing countries in the changing scenario. In this study we have made an attempt to trace the dynamics of marine exports from Kerala, giving special emphasis to changes in the present quality standards followed by different countries and the trends in the marine products exports. Export Inspection Council established its network throughout India so that the requirements of the importing countries are addressed in a timely manner and the need of the exporters is fulfilled without delays so as to facilitate exports. The present study is an attempt to capture the development in the seafood export industry in the world trade scenario and the overall impact and performance of marine exports. The Main trust of this study is to analyze the development and problems of fish processing and marine exports from Kerala.

Summary

Over the last ten years, India's merchandise trade has increased manifold from US $ 195.1 billion in 2004-05 to US $ 764.6 billion in 2013-14. As per the WTO, India's share in global exports and imports increased from 0.8 percent and 1.0 percent respectively in 2004 to 0.7 percent and 2.5 percent in 2013. It's ranking in terms of leading exporters and importers increased from 30 and 23 in 2004 to 19 and 12 respectively in 2013. Fisheries constitute about 1 percent of the GDP of the country and 4.75 percent of agriculture GDP. The total fish production during 2013-14 was 9.58 metric tonnes, an increase of 5.96 percent over 2012-13. Fish production during the first two quarters of 2014-15 has also showed an increasing trend and is estimated at 4.37 metric tonnes (Economic Survey, 2014-15).

Fish is generally accepted as the ensuing answer to the problem of protein-calorie malnutrition faced by the fast growing population of the world. Seafood is highly perishable in nature and is subject to spoilage unless handled, processed, packed and stored scientifically. It requires different infrastructure for handling, value-addition, processing and marketing. In 2012, people consumed about 136.2 million tonnes of fish. In the last five decades, world fish food supply increased at an average annual rate of 3.2 percent, outpacing global population growth at 1.6 percent. World per capita fish consumption increased from an average of 9.9 kg in the 1960s to 19.2 kg in 2012 due to population growth, rising incomes, urbanization, expansion of fish production and due to more efficient distribution channels (FAO, 2014). Stimulated by higher demand for fish, world fisheries and aquaculture production is projected to reach about 172 million tonnes in 2021. Estimates of 2011 indicate that exports of fish and fishery products exceeded US$ 125 billion, with average prices increasing by more than 12 percent (FAO, 2012). India is the second largest producer of fish in the world contributing to 5.43 percent of global fish production. India is also a major producer of fish through aquaculture and ranks second in the world after China. The consumption of fish has grown faster than that of any other animal products. The domestic demand for fish by 2015 has been projected as 6.7-7.7 million tonnes. According to the National Sample Survey (NSS), the annual per capita fish consumption was 2.45 kg in 1983. It increased to 3.45 kg in 1999-2000. Only 35 percent population in India was estimated to be fish eaters and their annual per capita fish consumption was 9.8 kg in 1999-2000. The per capita consumption of fish in India has increased to 10-11 kg/annum. The annual per capita consumption of fish in Kerala is 18.5 kg (Fisheries Dept.2012). An average Keralite consumes 27 kg of fish in a year and this tops even the global average of 19 kg. The per capita national average is way below at 9 kg per year (CMFRI, 2013).

The growth of export of marine fish products from developing countries have been regularly threatened by various non-tariff barriers from the developed countries. Non-tariff barriers relates to hygiene and food safety, in particular sanitary, traceability and quality control aspects. In order to cope up with the International quality standards, the challenges faced by export processing firms are to develop innovative technologies and efficient management to raise productivity to meet the growing demand for fish and fishery products at the lowest cost. An effective technology transfer inventions and innovation would play a crucial role in confronting a number of supply side obstructions and numerous demand side opportunities. It would greatly help in bridging the wide gap between the potential and the realized productivity. And the opportunities would be in augmenting income, generating foreign exchange earnings, productive employment creation and involving a number of additional beneficiaries in the value chain of fisheries sector.

The study also analysed the impact of Anti-dumping laws under the WTO framework and its impact on marine exports. The United States International Trade Commission (USITC) determined that the US industry is neither materially injured nor threatened with material injury by reason of imports of frozen warm water shrimp from India, China, Ecuador, Malaysia, and Vietnam. USITC voted 4-2 against imposition of countervailing duty (CVD) against India and other six countries. As a result of the USITC's negative determinations, US Commerce will not issue countervailing duty orders on imports of these products from India, China,

Ecuador, Malaysia, and Vietnam. The final decision of USITC brings great relief to Indian shrimp industry and its exports.

This study was taken up on a highly relevant background of the prospective growth of the marine export industry in Kerala. To gear up and meet the future challenges, as well as competition from the global level seafood industry, the Kerala seafood export processing industry are required to grow in size and volume.

Growth of Fish Production and Export Trends

During 2012- 13, export aggregated to 9, 28,215 tonnes in volume and valued at Rs 18,856.26 crore recording a growth of 13.6percent over previous fiscal. The historical scenario of Indian fisheries revealed a paradigm shift from marine dominated fisheries to a scenario where inland fisheries has emerged as a major contributor to the overall fish production in the country. Within inland fisheries, there is a shift from capture fisheries to aquaculture during the last two and a half decade. Freshwater aquaculture with a share of 34 percent in inland fisheries in mid-1980s has increased to about 80 percent in recent years. It has emerged as a major fish producing system in India as a result of the governmental initiatives taken during the past three decades. Fish Farmers Development Agencies (FFDAs) were set up in each district for delivering a package of technologies, practices, training and extension and for providing financial assistance to the beneficiaries. So far 429 FFDAs functioning in the country have brought about 0.65 million ha of water area under fish farming and reached out to 1.1 million beneficiaries and imparted training to about 0.8 million. Currently, the average annual yield is around 2.9 tonnes/ha. The Department has continued to extend financial assistance for the development of marine sector and for improving the socio-economic conditions of the traditional fishermen. The scheme has three major components, viz. Development of Marine Fisheries, Development of Infrastructure and Post Harvest Operations and Innovative activities. During 2012-13, 2,370 crafts were motorized, 8 minor fishing harbours and one fish landing centre was constructed, 27 infrastructure and marketing projects were approved and 1,400 safety kits were distributed to fishermen. During 2013-14, 6,260 crafts were motorized, 14 fish landing centers and 6 minor harbours was approved for construction, 27 infrastructures and marketing facilities were approved and 500 safety kits were distributed to fishers.

The Meenakumari committee recommendations on deep sea fishing has put traditional fishermen and mechanised fishing boat owners on the war path. The fisheries sector, of late, is seeing a series of protests against the move to implement the report, which they feel will affect their livelihood. Among other things, the committee had recommended creation of offshore buffer zones up to 200 meters, allowing foreign trawlers to fish beyond 12 nautical miles, legislation to regularise Indian fishing fleets, etc. The Central Marine Fisheries Research Institute commented that some of the findings of the committee are objectionable, including the creation of buffer zones. Traditional fishermen, already badly hit by the deep sea trawlers owned by foreigners and the aquaculture farms springing up in the country, will suffer a further blow. They fear that large-scale import of prawns, crabs, and cattle fish would lead to a drop in domestic prices and push fishermen into penury.

Over 55 varieties of marine products are exported to different countries in South East Asia, Europe, China, Japan and USA. The shrimp exports increased by 13 percent in quantity, 35 percent in rupee value and 41 percent in dollar value. The unit value realisation went up by 24 percent. The exports of value-added products showed a growth of about 2 percent in quantity and 12 percent in dollar realisation. The other major items of export are Frozen Fin fish, frozen cuttlefish, etc. Tuna is the third major fish commodity traded internationally after shrimp and ground fish. It constitutes 9 percent of the international trade in terms of value.

Catch utilization channelized into 80 percent of the fish catch is marketed as fresh or chilled and forms a staple food in the coastal and inland areas while 6 percent goes for drying and curing. Frozen fish production accounts for about 7 percent, while 6 percent is reduced to fish meal, and one percent used for other purposes. The fish canning industry utilises less than one percent of the total catch.

Seafood export processing industry growth is sustained through value addition, productivity accelerated through operational efficiencies and technical skills acquired. Export growth strategies should revolve around by expanding market to other geographical market segments and the creation of profit maximization and risk as well as cost minimization. Growth of the fish export processing industry in Kerala is financially evaluated on basis of economic performance and the total quality management. The result suggests that FDI have significant positive effect on export competitiveness of food processing industry of India. Seafood processing industry is one of the manufacturing industries in Kerala which can increase foreign exchange reserves through export. The demand for fish in future will be more than double as there is change in attitude of the consumer who prefers seafood as health food.

International Quality Assurance Standards on Fish

Most of the seafood export processing companies in Kerala have surplus installed capacities and they have to utilize their capacity to diversify production, create value addition products and must explore destination of new markets. Current issues are the quality of fish that should be maintained at a very high level right from capture by adopting various techniques. Government has to implement and direct processed quality fish products from catch to table in both domestic and international market. Lack of compliances with international standards and regulation could have a serious impact on international trading of aquaculture products from developing countries like India. Export inspection council of India (EIC) as an official certification body of India, as also the advisory body to the Government of India on matters of quality control and inspection, has implemented major changes to deal effectively with the recent trends in quality assurance especially in seafood sector which includes a major portion of its certification activity for agricultural products. In seafood, the major safety problems are microbiological, chemical and also related to the presence of antibiotic/veterinary drug residues. Spoilage and presence of certain muddy smelling compounds are the quality problems. To meet the requirements of international food trade, countries have formulated Quality standard to ensure food safety. These standards generally are in agreement with the conditions/stipulations of International quality standards i.e,

USFDA and EU standards. Most of the exporting countries adopt these standards for reasons of uniformity and international acceptability.

The international trading regulations are dynamic in nature due to increasing stringent quality measures of EU, US and other developed countries. Seafood safety regulations has resulted in more detention and rejections of fish and fishery products from developed countries and thereby led to widening of market destinations and increased market access to other countries. Both SPS and TBT agreements have relevance in the seafood industry and have given a new direction to the international food trade. The principles of achieving harmonization of standards and equivalency in food control systems and the use of scientifically-based standards are embodied in the two binding regulations of WTO. The main purpose of TBT agreement is to encourage the use of International standards, guidelines and recommendations in Technical regulation and standards, testing, packaging, marketing and labeling. Stringent international quality standards and regulations demands Kerala seafood industry to become more competitive. This study tried to capture demand side constraints include those related to international trade such as non-tariff barriers of quality standards and its impact on export market.

The work of the Codex Alimentarius Commission, together with that of FAO and WHO in their supportive roles, has provided a focal point for food-related scientific research and investigation, and the Commission itself has become an important international medium for the exchange of scientific information about the safety of food. The standards of Codex have also proved an important reference point for the dispute settlement mechanism of the WTO. Quality related problems have become dominant in the seafood export processing industry in Kerala. Seafood export processing industry in Kerala has to operate with higher cost to ensure compliance with new requirements in the global market. Seafood quality is affected by handling, gutting, processing and storage temperature. Quality is one of the aspects which are constantly changing with the improvement of technology and innovations. As and when importing countries introduce new regulation, seafood exporter must be cautious to new regulations and comply with them in addition to ensuring the implementation for the betterment of the export. A widespread educational and awareness generation programme on quality standards should be a necessary condition for solving quality issues. Unless we achieve both productivity and quality revolutions, our farm products will not be globally competitive.

U S Anti-Dumping Duty on Frozen shrimp Exports

India is one of the highest users of anti-dumping and is second only to United States in the year 2001, according to the WTO sources. India is also one of the main victims of anti-dumping action by foreign authorities. India has been affected by the antidumping duties levied on certain frozen shrimp exported to the US, based on a petition filed by the Ad Hoc Shrimp Trade Action Committee in the US to the Department of Commerce in the US. The investigations proved to be a major setback for the Indian shrimp exporters. There was significant fall in export to the US. Another major impact that the AD investigations and the subsequent imposition of the AD duties on India had been resulted in a fall in the number of

Indian exporters exporting shrimp to the USA. The duty along with the continuous bond requirement, equivalent to the value of product exported, proved to be a major setback to the exporters. The use of anti-dumping measures as an instrument of fair competition is permitted by WTO. Antidumping laws are the world's most dominant form of trade protectionism, against their trading partners. The NTBs cover a wide spectrum and includes SPS, TBT, AD and Subsidies and Countervailing measures. The best way to evade the AD duty is product differentiation with value added products like ready to pan or ready to eat. This will surely enrich fish processing exporters to capture greater gains from trade and wider market access which will automatically develop brand name for our products. Recent records showed that around 5 percent of Kerala's shrimp exports are value added products. Rightly this is the time for us to have a shift in our production to finished products and come out the vicious circle of raw product to finished products. U.S. antidumping law is a tangled and confusing subject because U.S. law and procedures have changed substantially over time. And due to this Indian Frozen Shrimp exporters foresee a market diversification and product differentiation challenges in order to attain better exporter earnings. Other issues like eco-labeling, country of origin, traceability, sustainability, appropriate marketing strategies should be explored. This should be implemented with the help of public private partnership formulation and with the help of required policy allusion.

Production Constraints on Fish Export Processing

Fish and Fishery export processing industry are facing lot of difficulties at the time of producing and packaging and exporting in the current period. Marine export industry's productivity and efficiency can be increased so that it can meet the increasing consumer's demands as well as can improve its competitiveness in the international market. Financial and technical support should be governed by government of India to enhance the productivity and efficiency in the food processing industries. The investment by public private partnership should be encouraged so that it can increase the competitiveness of the products by reducing the cost apart from the stringent SPS regulations.

Marketing access is one of the vital areas where fish and fishery `export processing industry confront problems including product differentiation, brand building, lack of market knowledge, availability of a growing domestic market, Inadequate infrastructure facilities, including power, water, roads, etc. and the complexities of international trade. Seafood industry is labour intensive one and it spreads throughout the entire coastal area of our country. The industry must place more thrust on new markets and new products. The industry attempts to make more value added products. For healthy and abundant fishery resources, exporters must improve management and decrease the political and economic pressures that lead to over fishing and depletion. High transaction cost and procedural delay lead to high fixed costs.

The major import of fish and fish products into India are fresh fish, chilled fish, and feed from aquatic products and are mainly imported from countries such as Bangladesh, Japan and Pakistan, Norway, China, Singapore, Thailand and

Republic of Korea. Fish feed is imported from countries like Thailand, Chile, Peru, Myanmar and Taiwan. The Ministry of Agriculture issued a notification on 7 July 2001, which requires animal products importers to produce a sanitary permit at the customs gate before entry into the country.

The seafood processing export industry are facing the problems of complicated exporting procedures, high shipping costs, cut-throat competition in the industry, changing quality standards of importing countries, irregular power and raw material supply, hygiene problems and non availability of quick transportation facilities from the fishing port to the processing units.

Kerala is rich in fishery resources as well as it possesses fish workers more skilled than other states. Traditional workers are aged more than fifty and their successors are not willing to follow the seafood processing workers as it is more time consuming and harder than other works due to the cold temperature maintenance. Moreover, people in Kerala are comparatively much educated than other states and the young population are not willing to work in the area of seafood processing export industry because their opportunity cost is much higher in other job areas. In Kerala, age group of more than forty suffers from health problems. Majority of the fish processing workers in Kerala are from the Northern States of country and they have limited demands than the Keralites, and are not bothered about their rights and are willing to work at low wages.

The use of anti-dumping measures as an instrument of fair competition is permitted by WTO. Antidumping laws are the world's most dominant form of trade protectionism, against their trading partners. The NTBs cover a wide spectrum and includes SPS, TBT, AD and Subsidies and Countervailing measures. The best way to evade the AD duty is product differentiation with value added products like ready to pan or ready to eat. This will surely enrich fish processing exporters to capture greater gains from trade and wider market access which will automatically develop brand name for our products. Recent records showed that around 5 percent of Kerala's shrimp exports are value added products. U.S. antidumping law is a tangled and confusing subject because U.S. law and procedures have changed substantially over time. And due to this Indian Frozen Shrimp exporters foresee a market diversification and product differentiation challenges in order to attain better exporter earnings.

India's Foreign Trade Policy 2009-2014, including a scheme called Vishesh Krishi Gram Upaj Yojana aimed at boosting exports of marine products fruits, vegetables, flowers, some forest products, and related value-added products. Under the plan, exports of these items qualify for a duty-free credit that is equivalent to 5 percent of their free-on-board export value. The credit is freely transferable and can be used to import a variety of inputs and capital goods. In order to solve the over-investment in fishing and processing capacity, MPEDA has drawn up various market promotion programmes for projecting our resource potential, product diversification, quality assurance and liberal incentives for investments and joint ventures.

Findings of the Study

1. The total value of marine products export from India was Rs.30213.26 crore in 2013-14 which accounts for approximately 1 percent of the total export from India. The quantity of marine exports from India during 2013-14 was 983756 tonnes. When compared to the previous year, it recorded a growth of 60.23 percent in rupee terms, 5.98 percent in volume and 42.60 percent growth in US$. During 2013-14, the export earnings have crossed 5 billion US dollars. Frozen shrimp is continued to be the single largest item of export in terms of value. It accounts 64.12 percent of the total U S $ export earnings.
2. India is currently the 18th largest export market for U.S. goods. The stock of U.S. foreign direct investment (FDI) in India was $24.7 billion in 2011 (latest data available), down from $24.8 billion in 2010. U.S. FDI in India is largely in the professional, scientific, and technical services, finance/ insurance services, and the information services sectors.
3. The major food safety problems encountered for fish and fish products are microbiological contaminants due to lack of hygiene in the production process and problems due to temperature changes caused by a broken cold chain. Prohibited use of antibiotics has caused major problems in recent years. Antibiotics are available for the treatment of diseases in shrimp and often are used in fish farms in China, India, Thailand, and Vietnam. However chloramphenicol and nitro furan are prohibited in Europe and consequently, a zero level tolerance has been enforced.
4. There has been significant market diversification in India's trade in recent years a process that has helped in coping with the sluggish global demand, which owes to a great extent, weakens in the Euro Zone. Region wise India's export shares to Europe and America have declined over the years from 23.06 percent and 20.1 percent respectively in 2004-05 to 18.6 percent and 17.2 percent respectively in 2013-14.Conversely the shares of India's exports to Asia and Africa have increased from 47.9 percent and 6.7 percent respectively in 2004-05 to 49.9 percent and 6.7 percent respectively in 2013-14. The change in direction immediately prior to the global financial crisis and since 2010-11 indicates the process of diversification underway.
5. A comparison of India's trade in the pre-crisis (2004-05 to 2007-08) and post crisis (2010-11 to 2013-14) shows that India's exports and imports from Europe, USA, Singapore have declined while its trade with Asia and Africa has increased.
6. In 2014-15 India's export to European region grew by only 0.2 percent. India's export to Africa and America grew by 12.9 and 14.4 percent respectively and to Asia, a major destination accounting for nearly 50 percent of India's export, by 2.2 percent in 2014-15(April-December).Within Asia India's exports to South Asia grew by 23.8 percent mainly due to high export growth to Sri Lanka, Nepal and Bangladesh) and 8.8 percent in the case of West Asia-Gulf Cooperation Council (GCC) (UAE, Saudi Arabia, and others) .India's exports to other regions of Asia witnessed a contraction

declining by 4.4 percent to North East Asia consisting of China, Hong Kong, Japan, 7.2 percent to the Association of South East Asian Nations(ASEAN) consisting of Singapore, Indonesia, Thailand, Malaysia and 8.5 percent to Other West Asian Countries (Iran, Israel and Others) in 2014-15.

7. Fish industry faces many challenges. One of the issues is to maintain the quality of fish at a very high level right from capture by adopting various techniques. Bio-terrorism, ecolabelling, traceability and risk assessment are some issues to be addressed. Accessing international seafood market has become complicated by the need to comply with the regulation concerning, product quality and safety requirements. Lack of compliances with these standards and regulation could have a serious impact on international trading of aquaculture products from developing countries like India.

8. Recent trends in marine products exports include an increased emphasis on food safety regulations in international trade, a tightening of standards by major importing counties, a reorientation of private sector, quality assurance, technique towards preventive management and the corresponding shifts in the role of EIC toward processed based including mandatory HACCP. Almost 200 countries supply fish and seafood products to the global market place and constant monitoring and enforcement is required to ensure safety of food. In seafood, the major safety problems are microbiological, chemical and also related to the presence of antibiotic/ veterinary drug residues. Spoilage and presence of certain muddy smelling compounds are the quality problems.

9. There were 179 exporters in 2001-02 and by 2004-05 the number had fallen to 109. The duty along with the continuous bond requirement, equivalent to the value of product exported, proved to be a major setback to the exporters. Imposition of 10.16 percent anti-dumping duty cast a cascading effect on fishing and aquaculture industries in the country. India continues to be on the top of the list with 39 safeguard initiations out of the world total 295 initiations and Indonesia ranks second with 26 safeguard initiations for the period from 1995 to 2014.

10. Fisheries scenario has been categorized into four Phases which examines the profile, the production, export trends, destination changes, product diversification, technological innovations, recent challenges and issues explored. During the I Phase, Japan, the EU and the US import a major proportion of India's exports and the main export species was shrimp. The compound annual growth rate of marine export quantity during the phase I from 1960 to 1985 was 9.2. The year 1985-86 has seen to boost economic growth, reinforce anti-poverty programmes, revitalize industry and provide a rapid expansion path to fisheries trade. The Indian Seafood export has shown an increasing trend during the II Phase. During the 1990s economic liberalization gave the fishing industry a major investment boost. The compound annual growth rate of marine export quantity during the II phase from 1985 to 1997 was 16. The third phase embarks upon the structural changes of seafood export processing industry that took place

in India especially in Kerala. The main hindrance for export in the third phase was the ban period due to quality issues. Stringent Quality Standards followed by EU, USA, Japan and the implementation of HACCP in seafood processing plants became mandatory. The compound annual growth rate of marine export quantity during the III phase from 1997 to 2004 was 4.7. The major barriers in the seafood export market in the fourth phase have been in the form of sanitary and phyto-sanitary measures imposed by the developed countries and the anti-dumping duty of frozen shrimp imposed by US. According to the Economic Survey 2012-2013, growth in exports can only be achieved with greater diversification of products. India has been fairly successful in diversifying its export markets from developed countries like the US and Europe to Asia and Africa, which has helped to a great extent in weathering the global crisis of 2008. Compound annual growth rate of marine export quantity during the IV Phase from 2004 to 2014 is 8.8. Fisheries sector witnessed a steady growth and the overall compound annual growth rate of marine export quantity from 1960 to 2014 is 8.4.

11. The fishing industry in Kerala is of vital importance in the economic development of the State. Starting from a purely traditional activity in the fifties, both aquaculture and fisheries have transformed to commercial enterprises and have considerable potential for employment generation and contribution to the food and nutrition security and foreign exchange earnings of the country. Fish production has a crucial role for the development of marine export processing industry. Compound annual growth rate of total fish production between 1950 to 2014 is 5.2 percent. Fish production levels have increased from 2.8 million tonnes of fish and shell-fish in 1985-86 to 5.3 million tonnes in 1996-1997. Compound annual growth rate of fish production during the I Phase upto 1985 is 10.9. The compound annual growth rate of fish production during the II Phase i.e. between 1985 to 1997 was 6.3. Inspite of sizeable investment for marine fisheries in the past, the growth in marine fish production in 1990's has been rather slow at an average of 2.1 percent when compared to the inland fish production growth rate of 6.5 percent. Compound annual growth rate of fish production during the III Phase i.e. between 1997 to 2004 was 2.8. The main source of growth is from aquaculture production because productions in capture fisheries are stagnant since mid-1990s due to depletion of fishery resources. Compound Annual Growth rate of fish production during the IV Phase 2004 to 2014 is 4.7.

12. The regression estimates for Indian marine exports show that the variables such as Thailand shrimp export value to USA(TEUS), Dummy Variable (D) are found to be statistically significant when compared to Indian Shrimp Export to USA (Value in Crores(IUS). The R^2 and adjusted R^2 value are also high. Regression estimates show that WTO rules and regulations and Anti-dumping duty on frozen shrimp by USA significantly influence Indian shrimp export to US.

13. Gravity Model shows that the growth in GDP has a positive impact on Marine fisheries export trend from India. In the model, the coefficient of the GDP is positive and highly significant. The distance variable is significant and has the anticipated negative sign, which indicates that India tends to trade, with its neighboring countries during the period of food safety regulations. This variable is expected to have a negative effect on trade, as transport cost increase with the distance between countries. Log GDPj of importing countries is significant at one percent level. However LogGDPi of India is significant only at 10 percent level. The value of R^2 and adjusted R^2 are found to be high.
14. The result of the gravity model with respect to USA reveals that growth in GDP of India and USA has a positive influence on the marine fisheries export from India. Log GDPj of USA is significant at one percent levels. However Log GDPi of India is significant only at 10 percent level. LogDij shows the right sign. However it is significant only at 10 percent level. The value of R^2 and adjusted R^2 are found to be high.
15. The result of the gravity model with respect to European Union reveals that growth in GDP of India and European Union has a positive influence on the marine fisheries export from India. Log GDPj of European Union is significant at one percent level. However LogGDPi of India is significant only at 10 percent level. LogDij shows the right sign. However it is significant only at 10 percent level. The value of R^2 and adjusted R^2 are found to be high.
16. The result of the gravity model with respect to Japan reveals that growth in GDP of India and Japan has a positive influence on the marine fisheries export from India. Log GDPj of Japan is significant at one percent level. Similarly LogGDPi of India is significant 1 percent level. The value of R^2 and adjusted R^2 are found to be high.
17. The Regression Analysis examines the effect of major trade barriers viz. international quality standards maximum residual limits and Anti-dumping duty on the marine product exports from India. The regression results reveal that the variables such as, Cadmium MRLs (CADj), Mercury MRLs (MERj), lead MRLs (LEADj), Anti-dumping duty dummy Variable (Anti Dj) is statistically significant. Total Marine Export Quantity (Tij) is having an inverse relationship Anti-dumping duty dummy Variable (Anti Dj).
18. The result of ANOVA proved that there are significant differences in the mean total cost of seafood export processing industry based on their installed capacity. Fisher's LSD (Least Significant Different) test is appropriate when we have 3 means to compare. If the significant value is greater than 0.05 then there is no statistically significant difference between two conditions that are being compared. In the case of our survey between 25 and 50, above 50 installed capacity firms are not statistically different in cost compliance to attain international standards. If the significant value is

less than or equal to 0.05, it indicates that there is a statistically significant difference between the two conditions being compared. Here in the survey group below 25(between 25 and 50, above 50) shows statistically significant difference in the cost incurred by the surveyed industry in complying with seafood safety and other standards in major import markets.

19. The result of ANOVA proved that there are significant differences in the mean total capacity utilization of seafood export processing industry based on their installed capacity. LSD-Post Hoc Test proved that If the significant value is greater than 0.05, for units between 25 and 50 and above 50, it indicates that those firms are not statistically different in their capacity utilization. If the significant value is less than or equal to 0.05 it shows that there is a statistically significant difference between the two conditions being compared and is shown in the mean difference column. The survey group below 25 (between 25 and 50, above 50) shows statistically significant difference in the capacity utilization from the surveyed units.
20. Survey observed that shrimp price spread is greater than other varieties of fishery products and at the same time PSFCR share is greater in shrimp as its demand is high in the importing countries. Fish products have higher PMMCR and this shows huge margin taken by intermediaries and indicates less efficiency in the supply chain system.
21. Out of the seven parameters which represent the problems and issues of seafood processing industry the major problem identified though Kruskal H static non-parametric test is quality issues. The second key problem recognized as raw materials and the third problem investigated as finance. The Fourth problem identified as Labour and fifth as Administrative and Technical issues. The sixth rank identified as production and the least rank is given to issues related to Marketing.
22. Price fluctuation may be one of the sources of exploitation among the supply chain actors and may lead to an artificial scarcity of raw materials. Shortage of trained labours is the core concern faced by seafood industry. The major issue in Marketing is the changes in the market trends and the competition from the branded firms. The main issues in the finance sector are the high interest rates and the delayed official procedures which definitely affect the rolling of cash in the seafood business. The administrative and technical department confronts the main problem as the changes in the international quality standards. They should always be updated on the changing standards and should be able to prepare relevant certification and documentation accordingly Another problem is the quality issue. Power scarcity and technological upgradation are also other issues facing the industry.
23. Kruskal H static non-parametric test proved that the foremost problem identified by the surveyed units is corruption practices pervading in Commercial Procedures and documentation and Coordination activities of Pre-shipment Clearance. The study suggested that all document should

be minimized and shipping should be permitted on invoice basis which involves all requisite information. In order to overcome the problem of corruption, Government should reduce the discretionary powers of the concerned officials.

24. Kruskal H static non-parametric test identified that the foreign exchange rates and prices are affecting exports. Changes in the dollar rates partially affected the prices of seafood products during the global recession. Another problem factor is the demand for marine fish products. For example, European Union countries, Italy and Germany demanded cheaper fishery products than high rated fishery products and is revealed in the responses of surveyed units.

Recommendations and Policy Implications

1. Government has an important role to play in achieving the country's potential for fish production. It is worth noting that certain European countries have imposed strict non-tariff restrictions on export of fish and fishery products. Several processing units that could not meet the European Union standards have been treated as sick units. The central government should re-examine the policy as it would affect the survival of small scale marine export processing units.
2. Seafood related diseases caused by bacterial pathogenic or their toxins are associated with improper hygiene standards, handling and storage. In order to reduce the number of seafood related outbreaks, coordinated efforts by several agencies are required.
3. Studies show that most of the increase in cross contamination in developing countries like India is due to the problems in maintaining the temperature stability. Minimum prescribed standards of fish markets and transportation facilities are to be implemented in order to follow international quality standards.
4. Other issues like eco-labeling, country of origin, traceability, consumer awareness, health environment and sustainability and suitable marketing strategies should be implemented with the help of public private partnership to smoothen the fish and fishery exports.
5. Exporters are looking for new export promotion schemes to export goods especially unification of raw materials prices than auction to avoid the indulgence of intermediaries in the supply chain of seafood industry. This should help exporters to access raw materials at a cheaper rate so that Indian products can become competitive in the international market.
6. The Maritime Stewardship Council Certification must be made compulsory and all fish and fishery export processing units should follow the rules and regulation to avoid depletion of fishery resources and the development of sustainability.

7. Marketing access is one of the crucial areas where fish and fishery export processing industry is facing problems and it include product differentiation, brand building, lack of market knowledge, availability of a growing domestic market, Inadequate infrastructure facilities and other complexities of international trade.
8. High transaction costs and procedural delays lead to high fixed costs. To solve this problem export documentation should be transparent and a single window system should be opened to avoid delays and energy waste.
9. Waste from processing factories can also be converted into useful food and non-food products. Crustacean processing waste can be made into new products like chitin and chitosan. Shark cartilage, squalene etc. have affiliations in medicine and cosmetics. Public Private Participation should be considered to solve this problem seriously and a suitable solution to reduce wastage of resources and creation of productive employment opportunities should be implemented.

Contributions of the Study

The Study identified some of the important quality issues and impact of anti-dumping duty on fish and fishery export processing industry in Kerala. The extensive survey and secondary data analysis made by the researcher will give more insights about the problems and economics performance of marine products exports from Kerala. The present investigation examined the non-tariff barriers faced by fishing industry in the international market and other related issues. The study analyzed the changes in the demand for consumption of fish and fishery products at the domestic as well as international markets. It is a very dynamic industry and the profitability is unpredictable. Sustaining competitive advantage depends on the availability of raw material with international quality and best market access strategies. The present study identified some of the problems related to Quality standards and the expenditure incurred and benefits acquired by the marine export processing industry. The study revealed that full capacity utilization is impossible due to non-availability of raw material. There is a limit to product diversification because the product is highly perishable. The findings are deemed to help the policy makers, researchers, exporters, quality controllers in India as well as Kerala.

Areas for Further Research

The present study explored the seafood export processing industry in Kerala and examined the impact of market service and market promotion activities. However the study can be extended to other states and districts in India. Besides this primary data can be extended to merchant exporters in Kerala. Further analysis and overall impact and performance of marine exports from Matsyafed-Kerala, State Co-operative Federation for Fisheries Development Limited and a comparative study of private and public limited companies of seafood export industry can be done by future researchers. Market services and market promotion have assumed special significance in view of the growing stiff competition from other seafood exporting countries in all overseas markets. Consequent on the large scale development of commercial scale shrimp farming in several shrimp producing countries in Asia and Latin America, the shrimp exporting countries are making all efforts to maximize share in all major world markets. These linkages and policies can be examined for stepping up of our promotional programmes in major overseas markets.

Conclusion

The present study made an in depth analysis of the marine products sector in Kerala. The study reveals that tariff barriers and non-tariff barriers plays an important role in the export of marine products from India. The non-tariff barriers cover a wide spectrum and include SPS, TBT, AD and Subsidies and Countervailing measures. Anti-dumping and countervailing measures are permitted to be taken by the WTO Agreements in specified situations to protect the domestic industry from serious injury arising from dumped or subsidized imports. Investing in the fishery sector is very important in reducing poverty and improving food security. Increased fish and fishery productivity generates a higher income and creates income-generating opportunities to come out from the vicious circle of poverty. The demand for fish in future will basically be determined by increase in the number of consumers and the preference for seafood. Unless we formulate exact standards for locally marketed fishery products it will be difficult for us to enforce regulations to import fish and fishery products. The findings of the study have significant policy implications, in the sense that it throws light into the present seafood export scenario and pinpoints the barriers to entry for export.

NOTES

CHAPTER 1

1. **NAMA:** It refers to all products not covered by the Agreement on Agriculture. It includes manufacturing products, fuels and mining products, fish and fish products, and forestry products. They are sometimes referred to as industrial products or manufactured goods.
2. **IUU Regulation**: COUNCIL REGULATION (EC) No. 1005/2008 of 29 September 2008 establishing a Community system to prevent, deter and eliminate illegal, unreported and unregulated fishing, amending Regulations (EEC) No. 2847/93, (EC) No. 1936/2001 and (EC) No. 601/2004 and repealing Regulations (EC) No. 1093/94 and (EC) No. 1447/1999
3. **Catch certificate:** This catch certificate template has similar content to the standard Dissostichus catch document form used by the Commission for the Conservation of Antarctic Marine Living Resources (CCAMLR) and the statistical document forms used by the Indian Ocean Tuna Commission (IOTC), International Commission for the Conservation of Atlantic Tunas (ICCAT), Inter-American Tropical Tuna Commission (IATTC), and Commission for the Conservation of Southern Bluefin Tuna (CCSBT).
4. **Net Exporters:** Country or territory whose value of exported goods is higher than its value of imported goods over a given period of time.
5. **Competitiveness:** It is the ability of a firm or industry to develop and maintain an edge over market rivals. This edge can be achieved by increasing efficiency, quality, and product differentiation or by influencing demand.
6. **Marine Stewardship Council:** It is an independent non-profit organization which sets a standard for sustainable fishing. When buyers choose to purchase MSC certified fish, well-managed fisheries are rewarded for sustainable practices.

7. **Countervailing duty:** It is a duty imposed in addition to the regular (general) import duty, in order to counteract or offset the subsidy and bounty paid to foreign export-manufacturers by their government as an incentive to export, that would reduce the cost of goods. Imposing a countervailing duty is the answer to unfair competition from subsidized foreign goods.
8. **Anti-dumping duty:** It is a duty imposed to offset the advantage gained by the foreign exporters when they sell their goods to an importing country at a price far lower than their domestic selling price or below cost. Dumping usually occurs from the oversupply of goods, which is often a result of overproduction, and from disposing of obsolete goods to other markets.
9. **Quantitative restrictions or Quotas:** They are important forms of NTBs. As a protective measure, quota is more effective than the tariff. The effects of quotas are more rigorous and arbitrary and they tend to distort international trade much more than tariff, by helping importers and exporters to acquire monopoly power.
10. **Trade distortion:** It is a policy that alters the amount of trade, up or down, from what it would otherwise be. Agricultural subsidies, even if not based on quantity of exports, are trade distorting unless they are paid independently of whether and how much farmers produce.

CHAPTER 2

1. Carp: They are various species of oily freshwater fish of the family Cyprinidae, a very large group of fish native to Europe and Asia.
2. The TCMP was aimed at providing marine diesel engines, a variety of nets and nylon threads, insulated iceboxes and van, etc. As part of TCMP, one gillnet boat, one trawler, one Danish Seiner and one big boat for dory fishing were introduced. Also, two ice factories were constructed–one at Vizhinjam and the other at Kayamkulam.
3. The FAO-TAP was initiated to provide training to fishermen on new fishing methods, develop appropriate crafts and gears, construct new fishing harbours, etc. The initial step for fishing harbours in Vizhinjam and Beypore has commenced. The Malabar fishermen al trained to operate purse-seine.
4. The Indo-Norwegian Project(INP) for the development of the fisheries community in the erstwhile State of Travancore-Cochin came into existence in January 1953 following a tripartite agreement signed in New Delhi between the United Nations, the Government of Norway, and the Government of India
5. Fishmeal is a nutrient-rich feed ingredient used primarily in diets for domestic animals (poultry, etc.), recently used for feeding ruminant animals and sometimes used as a high-quality organic fertilizer.
6. Fish oil can be extracted from the fatty tissue of fish or from the liver of fish. Fish oil contains a higher level of Vitamin A and Vitamin D which are essential for health.

7. Fish soluble is a brown, semi-viscous liquid, with a fishy but pleasant odour. The unidentified growth factors present in fish soluble helps in increasing the rate of growth. The fish soluble is rich in all the dissolved proteins, vitamins and amino acids
8. Satpati is one of the biggest fishing villages on the western coast of India. It is about 80 km north of Mumbai, located in the Palghar Taluka of District Palghar in Maharashtra. The main industry in Satpati is fishing, with large exports abroad.
9. An outboard motor is a propulsion system for boats, consisting of a self-contained unit that includes engine, gearbox and propeller or jet drive, designed to be affixed to the outside of the transom.
10. An exclusive economic zone (EEZ) is a sea zone prescribed by the United Nations Convention on the Law of the Sea over which a state has special rights regarding the exploration and use of marine resources, including energy production from water and wind.[1] It stretches from the baseline out to 200 nautical miles (nmi) from its coast
11. MSY is theoretically, the largest yield (or catch) that can be taken from a species' stock over an indefinite period.
12. Tsunami is a Japanese word made up of two characters. TSU meaning harbor and nami meaning wave. History of Tsunami dating back from 6100 BC to 26 December 2004 has been tracked with starting facts and figures. Oldest record of an Indian Tsunami is 326 B C near Indus.Tsunami is 326 B C near Indus. Delta, Kutch region in which the mighty Macedonian fleet of Alexander the Great was destroyed the earthquake followed by the Tsunami in 1524 Tsunami occurred near Dabhole Maharashtra.
13. ICTSD- International Centre for Trade and Sustainable Development.
14. Economic union is a type of trade bloc which is composed of a common market with a customs union. The participant countries have both common policies on product regulation, freedom of movement of goods, services and the factors of production (capital and labour) and a common external trade policy.

CHAPTER 3

1. Biogenic amines: They fare a class of compounds that are derived from amino acids, the building blocks of proteins. The most well known biogenic amine is histamine, which is produced by the body and which plays an important role in allergic reactions.
2. ISO/IEC 17025: Laboratories use ISO/IEC 17025 to implement a quality system aimed at improving their ability to consistently produce valid results. It is also the basis for accreditation from an accreditation body. Since the standard is about competence, accreditation is simply formal recognition of a demonstration of that competence. A prerequisite for a laboratory to become accredited is to have a documented quality management system.

The usual contents of the quality manual follow the outline of the ISO/IEC 17025 standard.

3. ISO 22000: It is an international standard that defines the requirements of a food safety management system covering all organizations in the food chain from "farm to fork", including inter-related organizations such as producers of equipment, packaging material, and cleaning agents, additives and ingredients.
4. Cryoprotectant : Is a substance used to protect biological tissue from freezing damage (i.e. that due to ice formation)
5. Mycotoxin: Any toxic substance produced by a fungus.
6. Thawed: To change from a frozen solid to a liquid by gradual warming.

CHAPTER 4

1. Administrative Reviews An annual investigation by the Department of Commerce to determine whether the actual amount of unfair trade was greater than or less than the duty rate. The Review may result in duties being increased, decreased, or remaining the same.
2. Anti-dumping Agreement: If a company exports a product at a price lower than the price it normally charges on its own home market, it is said to be "dumping" the product. The WTO Agreement does not regulate the actions of companies engaged in "dumping". Its focus is on how governments can or cannot react to dumping it disciplines anti-dumping actions, and it is often called the "Anti-dumping Agreement".
3. Antidumping Duties A tax imposed by the U.S. government on imports that have been found (1) to be sold for less than fair value, and (2) to cause injury to the domestic industry.
4. Dumping: The practice of selling a product into the U.S. market for less than fair/normal value or below the cost of production. Dumping occurs when the Products of one country are sold in another at "less than fair value" which is usually taken to mean production cost plus reasonable profit.
5. Countervailing duties (CVDs), also known as anti-subsidy duties, are trade import duties imposed under World Trade Organization (WTO) rules to neutralize the negative effects of subsidies. They are imposed after an investigation finds that a foreign country subsidizes its exports, injuring domestic producers in the importing country.
6. Dumping: The practice of selling a product into the U.S. market for less than fair/normal value or below the cost of production. Dumping occurs when the Products of one country are sold in another at "less than fair value" which is usually taken to mean production cost plus reasonable profit.
7. CDSOA The Continued Dumping and Subsidies Offset Act also called the "Byrd Amendment." The repealed U.S. law allows duties collected

on imports before October 2007 to be distributed to U.S. producers that supported trade actions

8. Commerce The Department of Commerce, which determines whether or not U.S. trade laws have been violated
9. Tariff Line: Categories of products with tariff
10. Bound and Applied Tariff: Bound rates are the maximum levels of tariff that can be applied and applied rates are the rates that are actually applied
11. Bound Line: Line (categories of products) for which maximum levels of tariffs are fixed.
12. The Southern Shrimp Alliance (SSA) is an organization of shrimp fishermen, shrimp processors, and other members of the domestic industry in the eight warm water shrimp producing states of Alabama, Florida, Georgia, Louisiana, Mississippi, North Carolina, South Carolina, and Texas. Founded in 2002, the SSA works to ensure the continued vitality and existence of the U.S. shrimp industry.
13. ICE The Department of Homeland Security's Immigration and Customs Enforcement, which helps identify illegal imports.
14. ITC The International Trade Commission, which determines whether a domestic industry is injured by unfair trade.
15. ICE The Department of Homeland Security's Immigration and Customs Enforcement, which helps identify illegal imports.
16. WTO The World Trade Organization, which determines whether actions to impose restrictions on trade by a country are consistent with international trade agreements

BIBLIOGRAPHY

Books

Armand V Figenbaum (1983): "Total *Quality Control*" Mc Graw Hill, Newyork pp 388.

Avadhani.V.A.(1998): "*International Finance Theory and Practice*", Himalaya Publishing House, Mumbai,Pp 165-174.

Bremner Allan H (2002): "*Safety and Quality Issues in Fish Processing*" Woodhead Publishing Limited, Cambridge, England, 507 pp.

Brink Lindsey and Daniel Ikenson (2007): "The *Rhetoric and Reality of US Anti-Dumping Law*" Academic Foundation, New Delhi, 208 pp.

Bo Sodersten(1990): "*International Economics*", Macmillian Publishing Ltd, Hong Kong.Pp167-185,1990

Cherunilam Francis (2008): "*International Economics*", Tata McGraw-Hill Publishing Company Limited, New Delhi Pp 249-265.

D Rajasenan (2005): "Resource *Based Industries in Kerala*" *Eds. Directorae of Publicaions and Public Relations,* Cochin University of Science and Technology, Kochi, pp 139-160.

Da-Wen Sun (2005): "*Thermal Food Processing: New Technologies and Quality Issues*" Taylor and Francis, New York, USA, 664 pp.

E Vivekanandan (2011): "Climate *Change and Indian Marine Fisheries*" Marine Fisheries Policy Brief- 3, CMFRI Special Publication No. 105, Kochi, 97pp.

Frank Asche (2006): "*Primary Industries Facing Global Markets: The Supply Chains and Markets for Norwegian Food and Forest Products*" Copenhagen Business School Press, DK, 477pp.

J Anderson (2003): "*The International Seafood*" Woodhead Publishing, Cambridge, England, 225pp.

K Gopakumar (1997): *"Biochemical Composition of Indian Food Fish"* Central Institute of Fisheries Technology, Kochi,44pp.

K Gopakumar(2002): *" TextBook of Fish Processing Technology" Directorate of information and Publications of Agricultrue,* Indian Council of Agricultural Research" New Delhi,481pp.

K K Vijayan, P Jayasankar and P Vijayagopal (2007): *" Indian Fisheries- A Progressive Outlook,* Eds.Central Marine Fisheries Research Institute, Kochi,203pp.

M Srinathm,Somy Kuriakose and K G Mini (2005): *" Methodology for the Estimation of Marine Fish Landings in India"* Fishery Resources Assessment Division, CMFRI Special Publication No. 86. Kochi, 56pp.

M K Mukundan and S Balasubramaniam (2009): *"HACCP Concept in Seafood Industry"* Central Institute of Fisheries Technology, Kochi, 121pp.

M K Mukundan and S Balasubramaniam (2011): "Seafood Quality Assurance" Central Institute of Fisheries Technology, Kochi, 175pp.

Mohan Joseph, M and Pillai, N G K, (2007): *"Status and Perspectives in Marine Fisheries Research in India"* Eds.CMFRI Diamond Jubilee Publication, Central Marine Fisheries Research Institute, Kochi, 404pp.

M.P Shrivastava (1992): *"Capaciy Utilisation and cost of production",* Deep and Deep publications, New Delhi, 250pp.

Margret Will and Doriz Guenther (2007): *"Food Quality and Safety Standards"* GTZ –Division 45, Germany, 162pp.

Paul Greenberg (2005): *"American Catch: The Fight for Our Local Seafood"* Penguin Press,USA,307pp.

P Surendren (2002): *"The Kerala Economy Growth and Survival"* Vrinda Publications (P) Ltd.,Delhi, 131pp.

Rogers, Everett M. (1983):"Diffusion of Innovations" The Free Press, New York, 447pp.

R Sathiadhas, R Narayanakumar, S ASwathy (2012): *"Marine Fish Marketing in India"* Central Marine Fisheries Research Institute, Kochi, 276 pp.

T.S Gopalakrishnan Iyer, M K kandoran, Mary Thomas, P T Mathew (2005): *"Quality Assurance in Seafood Processing"* Central Institute of Fisheries Technology, Kochi, 252pp.

U K Srivastava and S Vathsala (1989): "Agro-*Processing Strategy for Acceleration and Exports"* Eds. Oxford and IBH Publishing Co. Pvt. Ltd. New Delhi, CMA Monograph No. 135, pp110-163.

V N Pillai and N G Menon(2000): *" Marine Fisheries Research and Management"* Eds. Central Marine Fisheries Research Institute, Kochi, 914 pp.

V Sriramachandra Murty (1997): "Status *of Research in Marine Fisheries and Mariculture"* Eds. CMFRI Special Publication Number 67, Kochi, 35pp.

V Venugopal (2006): *"Seafood Processing Adding Value Through Quick Freezing,*

Retortable Packaging, and Cook –Chilling" CRC Press, Taylor and Francis Group, New York, London, 447pp.

Journals

Anders S M and J A Caswell (2009): "Standards as Barriers versus Standards as Catalyst: Assessing the Impact of HACCP Implementation on U.S. Seafood Imports" *American Journal of Agricultural Economics,* Vol.91, No.2.pp.310-321, May 2009.

Ayyappan S and M Krishnan (2005), "Offbeat attempts needed", *the Hindu Survey of Agriculture* 2005,pp.134-136.

Bekaert Geert, Harvey R Canpbell, Lundblad T Christian, Siegel Stephen(2011): " The European Union, The Euro, and Equity Market Integration" *Journal of Financial Economics* 109 (2013) 583-603

Biggs D Stephen and Clay J Edward (1981): "Sources of Innovation in Agricultural Technology" *World Development,* Volume 9, Issue 4, pp321-336.

B P Sarath Chandran, P K Sudarsan(2002): "India-Asean Free Trade Agreement Implications for Fisheies",*Economics and Political Weekly,* April 21,2012 Vol XLVII pp.16

Das Gupta (2002): "Shaping the WTO Anti-dumping Agreement: A Developing Country Perspective | " *Foreign Trade Review,* Vol 36,No 3 & 4.

Day Madan Mohan, Roehlano M Briones and Muffuzzudin Ahmed (2005) "Disaggregated Analysis of Fish Supply, Demand and Trade in Asia: Baseline Model and Estimation Strategy", *Aquaculture Economics and Management,* Vol.9,No.1-2,pp.113-140.

Geethanjali Nataraj and Paravakar Sahoo(2004): "Enlargement of EU: Effect on India's Trade" *Economic and Political Weekly,* Vol. 39, No. 19(May, 8-14,2004),pp.1872-1876.

Gebrehiwet, Kirsten (2007): "Quantifying the Trade Effect of Sanitary and Phytosanitary Regulations of OECD Countries on South African Food Export" *Agrekon,* Vol 46, No 1 March 2007.

Ekmekcioglu Ercan(2012):"The Benefits and Problems of International Trade in Context of Global Crisis",*China-USA Business Review,* ISSN 1537-1514 April,2012,Vol.II, No.4

Harikumar G. & Rajendran G (2007): "An Over View of Kerala Fisheries – with Particular Emphasis on Aquaculture", *IFP Souvenir* 2007.

Hazari R Bharat and Sgro M Pasquale (2002): "Comparative Advantage and Trade Policy", International Economics Finance and Trade, in EOLSS encyclopedia of life support systems", *EOLSS/UNESCO,* Paris, France. Vol I

Ho Huy Tuu and Svein Ottar Olsen (2013): "Marketing Barriers and Export Performance: A Strategy Categorization Approach in the Vietnamese Seafood Industry" *Asian Journal of Business Research,* Vol. 3, No. 1, 2013

Hussan F,Jeeva Charles and Prathap Sangeetha K(2012),'Economics of cost of compliance with HACCP in seafood export units and its limitations for applicability in domestic markets", Central Institute of Fisheries Technology, Cochin, India. *Indian J. Fish, 59(1) 141-145, 2012*

Jacob T, Rajendran V, Mahadevan Pillai, Andrews Joseph and Satyavan U K (1987): "An Appraisal of the Marine fisheries in Kerala" *Indian Council of Agricultural Research*, 40th Anniversary Celebrations of CMFRI Special Publication, Number 35.

Kose, M. Ayhan, Christopher Otrok, and Charles H. Whiteman (2003): "International Business Cycles: World, Region, and Country-Specific Factors." *American Economic Review*, 93(4): 1216-1239.

Jayasekhar Somasekharan, Harilal K. N, Parameswaran M.(2013), "Coping with the Standards Regime: Analyzing Export Competitiveness of Indian Seafood Industry" *Sustainable Agriculture Research*, Vol 2, No 1 ,2013.

Judith M. Dean, Robert Feinberg, and José E. Signoret,, Michael Ferrantino† and Rodney Ludema(2008): "Estimating the Price Effects of Non- Tariff Barriers", U S International Trade Commission, *Office of Economics Working Paper*, No. 2006-06-A(r).

K M Shajahan (2012): The European Union Regulation on IUU Fishing Impact on Developing Countries" *Economic & Political Weekly EPW* August 11, 2012 VOL XLVII No.32.

Long B Russell, "United States Law and the International Anti-Dumping Code" *American Bar Association*, Vol. 3, No. 3 (April, 1969), pp. 464-489

Marucheck, A., Greis, N., Mena, C. and Cai, L. (2011): "Product Safety and Security in the Global Supply Chain: Issues, Challenges and Research Opportunities", *Journal of Operations Management* 29: 707-720.

Maarten Smeets (2013): 'Trade Capacity Building in the WTO: Main Achievements since Doha and Key Challenges' 47 Journal of World Trade, Issue 5, pp. 1047–1090

Morrow J D, Margolies G R, Rowland J and Robert L J(1991): " Evidence That histamine is the causative town of scombroid fish poisoning" New England, Journal of Medicine, 324:716-720

MPEDA(2003): "Anti-Dumping" Article", *Statistics of Marine Products*,Kochi.

Nair, K V Somasekharan and Jayaprakash, A A (1983): "Clash between purse seine and artisanal fishermen at cochin" *Marine Fisheries Information Service Technical and Extension Series*, 49. pp. 14-16.

Patnaik Kumar Jagaish(1997): "India and the GATT: Origin Growth and Devlopment", *A.P.H. Publishing Corporation*, New Delhi,1997.

P R Krugman(1993): "Free Trade: A loss of (Theoretical) Nerve? The Narrow and Broad Argument for Free Trade" American *Economic Review*, Vol. 83 No.2.

Ray Edward John (1981)," The Determinants of Tariffs and Non-tariff Trade Restrictions in the United States", *Journal of Political Economy*, pp, 105-21.

Ramakrishna Korakandy(1994): "Technological change and the Development of Marine Fishing Industry in India-A case study of Kerala" *Daya Publishing House*, Delhi.pp269

Ramakrishnan Korakandy(2008): "Fisheries Development in India- the Political Economy of Unsustainable Development" *Kalpaz Publication*, Delhi, pp184.

R Sathiadhas, R Narayanakumar and N Ashwathy(2012): " Marine Fish Marketing in India", *Central Marine Fisheries Research Insititute*, Kochi,276pp.

Sarada C , T Ravisankar, M Krishnan and C Anandanarayanan(2006): " Indian Seafood Exports: Issues of Instability, Commodity Concentration and Geographical Spread" *Indian Journal of Agricultural Economics*, Volume 61,No.2.April-June.

Sathiadhas, R and Immanuel, Sheela and Laxminarayana, A and Krishnan, L and Noble, D and Jayan, K N (2003): "Institution village linkage programme Coastal Agro Ecosystem & Interventions" *CMFRI Special Publication*, 75. pp. 1-79.

Sathyan Naveen, V. V. Afsal , V Thomas Joice (2014): "The present status of sea food pre-processing facilities in Kerala with reference to Alleppey district", Network for Fish Quality Management & Sustainable Fishing (NETFISH), *International Journal of Research in Fisheries and Aquaculture*, 4(1): 39-46.

Silva De D.A.M. and Yamao Masahiro (2008): "Compliance on HACCP and export penetration: An empirical analysis of the seafood processing firms in Sri Lanka" Sabaramuwa *University Journal*, Volume 8 Number 1; December 2008, pp 61-77

Singh Jeet, Yadav Preeti (2009): " Impact of Global Recession on Indian Economy with special reference to India's Exports" *Management Insight* , Vol. V, No. 2; December, 2009

Shyam S S, Uma K and Rajesh S R (2004): "Export Performance of Indian Fisheries in the Context of Globalisation", *Economics and Department of Agricultural and Rural Management*, Tamil Nadu Agricultural University, Ind. Jll. of Agri. Econ. Vol. 59, No. 3, July-Sept. 2004

Sudsawasd Sasatra (2012):" Tariff Liberalization and the Rise of Anti-dumping Use: Empirical Evidence from Across World Regions" *The International Trade Journal*, Volume 26, Issue 1, 2012, pp 4-18.

Tharakan, P K M and J. Waelbroeck(1994): (994): "Anti-Dumping and Countervailing Duty Decision in the EC and in the US.: An experiment in comparative political economy", *European Economic Review* Volume 38, Issue 1, January 1994, Pages 171–193

Thomas J Prusa (2005): "Anti-dumping: A Growing Problem in International Trade" *World Economy*, Vol.28, Issue 5, pp683-700.

Maarten Smeets, 'Trade Capacity Building in the WTO: Main Achievements since Doha and Key Challenges' (2013) 47 *Journal of World Trade*, Issue 5, pp. 1047–1090

Saraswati (2014): "Export Potential of Food Processing Industry in India" *International Journal of Computing and Corporate Research*, Volume 4 Issue 2 March 2014.

Sathiadhas, R and Fameena Hassan, 2002, Product diversification and promotion of value added products with quality control in Indian Seafood Trade, *Seafood Export Journal*, xxxiii (8-9):27-42.

Salim.S.Shyam(2012): Indian Seafood Industry and Post WTO-A Policy Outlook Socio-Economic Evaluation and Technology Transfer Division, *Central Marine Fisheries Research Institute*, Cochin.

S.S. Shyam, C. Sekbar, K. Uma and S.R. Rajesh (2004): "Export Performance of Indian Fisheries in the Context of Globalisation" *Indian Journal of Agricultural Economics* Vol. 59, No. 3, july-Sept. 2004

Sawhney Aparna (2005): "Quality Measures in Food Trade: The Indian Experience", *Indian Institute of Management*, The World Economy, Vol. 28, No. 3, pp. 329-348, March 2005.

S Jayasekhar (2009): "Food safety regulations and trade effects: The Case of Indian Seafood Industry with special reference to Kerala, South India" Doctoral scholar, *Center for Development Studies*, Trivandrum, Kerala.

Petko Draganov(2012): "Non-tariff measures: A key issue in evolving trade policy "Deputy Secretary-General, *United Nations Conference on Trade and Development*, October 01, 2012

Trondsen, Torbjorn (2012): "Value chains, business conventions, and market adaptation: A comparative analysis of Norwegian and Icelandic fish exports" *Canadian Geographer*, winter 2012, Vol. 56 Issue 4, p459-473.

Yusuf, M. and Trondsen, T. (2013): "A market-oriented innovative quality framework for the investigation of competitive entry opportunities into new seafood markets for producers", *Int. J. Quality and Innovation*, Vol. 2, No. 2, pp.175–192

Thesis

Albin T (1992): "Diffusion of computer numerically controlled machines tool in India" M Phil Thesis, Centre for Development Studies, Thiruvananthapuram.

Ancy Sebastian (2009): "Development of Safety and Quality Management System in Shrimp farming" Thesis, School of Industrial Fisheries. CUSAT, Kochi.

A V Shibu (1999): "Comparative Efficiency of different types of mechanical fishing operations along the cochin coast" Thesis, Department of Industrial Fisheries, CUSAT, Kochi.

Babu V (2001): "Dynamics of pesticides in the backwaters of Kuttanad" Ph.D Thesis in Environmental Chemistry, School Environmental Studies, Cochin University Science and Technology, CUSAT, Kochi.

Beena Sulochanan (2008): "Effect of Natural Antioxidant Sources on the quality upgradation of processed products from Indian Mackerel" Thesis, School of Industrial Fisheries. CUSAT Kochi.

Chebet L. "Rapid" (alternative) methods for evaluation of fish freshness and quality, Msc Thesis, Akureyri, Iceland: University of Akureyri; 2010. http://skemman.is/stream/get/1946/5682/16909/1/Masters_thesis_2010_Lillian.pdf accessed on 5th January,2014

Dong Phuong Thi Khuu (2012): "The Impacts of Non-Tariff Barriers on the Export Price of Vietnamese Catfish" Thesis, Fisheries and Aquaculture Management and Economics,Nha Trang University, Vietnam, May 2012.

D Rajasenan(1987): " Fishing Industry in Kerala Problems and Potentials" Thesis, Cochin University Science and Technology, CUSAT, Kochi.

Gopalakrishnan K S, 1990): "Studies on the toxic effects of some pesticides on the fish"Thesis, Department of Industrial Fisheries, CUSAT, Kochi.

Francis C A (1991): Export Development: Process and Potential A Study with Special Reference to the Marine Products Industry in India, Thesis, Cochin University of Science and Technology, Cochin

Gopalakrishnan K S, 1990): "Studies on the toxic effects of some pesticides on the fish "Thesis, Department of Industrial Fisheries, CUSAT, Kochi.

George Bindu (2011): "Technological Change and Modernisation in the Fishing Sector: The Question of Sustainability, Department of Applied Economics, Thesis, Cochin University of Science and Technology, Kochi.

J D Reyes M.A (2010): "Trade Liberalisation and the Role of Non-Tariff Barriers to International Trade" Thesis, Georgetown University, Washington, DC.

Joice V Thomas(2003): "Impact of Trawling on sea bottom and its living communities along the in shore waters of south west coast of India" Thesis, School of Industrial Fisheries, Cochin University Science and Technology, CUSAT, Kochi.

John Kurien (1994): "Technology Diffusion in Marine Fisheries: The Concrete Socio Economic and Ecological Interrelations" Thesis, Tata Institute of Social Sciences, Bombay.

John A T (1996): "Energy Optimisation Studies in Trawling Operations along Kerala Coast" Thesis, CUSAT, India, pp 213.

Jacob P Jose, 1988): "Combined toxic effects of oil and pesticides in selected Marine invertebrates, Thesis, Department of Industrial Fisheries, CUSAT, kochi.

John Mohan (2007): "Studies on some aspects of Landings, Utilization and Export of commercially important Cephalopods, Ph. D Thesis, School of Industrial Fisheries, Cochin University Science and Technology, CUSAT, Kochi.

Loc Thanh Thi Vo (2006): "Seafood supply chain quality management: the shrimp supply chain quality improvement perspective of seafood companies in the Mekong Delta, Vietnam" Thesis. Centre for Development Studies, Netherlands.

Nguyen Thi Van Anh (2009): Effects of Food Safety Regulatory Standards on Seafood Exports to US, EU and Japan", Thesis, The Graduate Faculty of Auburn University, Alabama, August 10, 2009.

Ouseph. T.P. (2005): "Dynamics of Export promotion strategies and global trade environment of export performance: studies on disaggregated export demand function with reference to Kerala" Thesis, Mahatma Gandhi University, Kottayam.

Pace M Kathryn (2011): "EU Import Notifications as a Protectionist Move: An Examination of the Relation between Tariff and Non-Tariff Barriers in Seafood Trade" Thesis, University of Illinois, Champaign.

Paul M Shibu(1992): "Trace metal speciation in the cochin Estuary" Thesis, Department of Industrial Fisheries, CUSAT, Kochi.

P Parvathy(2012): "Impact of the Sanitary and Phyto Sanitary Agreement on the Export of Marine Products from India with Special reference to Kerala" Thesis, Cochin University of Science and Technology,Kochi.

Prafulla V (2002): " Investigations on the Distribution characteristics of heavy metals in Squid, in relation to the levels in food fishes from the west coast of India with a perspective of seafood safety" Thesis, Ph.D in Marine Science, Cochin University Science and Technology, CUSAT, Kochi.

P Vijayagopal(2003): " Nutritional responses in Indian white Shrimp to varying protein Energy combination in compounded artificial feeds" Thesis, School of Marine Science, CUSAT, Cochin.

Ramachandran A (1988): "Studies on the production Management in the Seafood Processing Industry in Kerala, Thesis, Cochin University Science and Technology, Cochin, India.

Reyes Daniel Jose(2010): "Trade Liberalisation and the Role of Non- Tariffs Barriers to International Trade" , Thesis, Graduate School of Arts and Sciences of Georgetown University Washington, DC August 24, 2010.

S Shassi (1998): "Development of Total Quality Management System in Seafood freezing industry in Kerala" Thesis, Department of Industrial Fisheries, CUSAT, kochi.

Sarkat Sinha Roy(2004): "Factors in the Determination of India's Exports" Thesis,Jawaharlal Nehru University, New Delhi.

Sebastian Mathew (1986): "Growth and Changing Structure of the Prawn export industry in Kerala 1953-83" Thesis, Centre for Development Studies, Thiruvananthapuram.

Smitha Nair(2007): "Market Orientation in Indian seafood Processing firms" Thesis, School of industrial Fisheries, CUSAT 2007.

Sreenath P G (2009): "Standardisation of process parameters for ready to eat fish products indigenous polymers coated Tin Free Steel Cans" Ph.D Thesis, Fish Processing Division, Central Institute of Fisheries, Technology ICAR, Kochi.

Verghese C P (1994): "Techno Economic Studies on Industrial Fishing in the Upper East Coast of India" Thesis, Department of Industrial Fisheries, CUSAT, kochi.

Valsala John (1977): "Marine Products Export Industry in Kerala. Some aspects of its structure and Backward Linkages" Thesis, Centre for Development Studies, CDS, Thiruvananthapuram.

Venkateswara Rao(1971) "An empirical study of us foreign trade sector according to standard international trade" Thesis, Delhi School of Economics, University of Delhi.

Ved Prakash (2004): "Anti-Dumping Countervailing Measures and Safeguard Provisions in Multilateral Trade Regime" Thesis, Delhi School of Economics, University of Delhi.

V M Abdul Hakkim (1980): "Mechanisation of Co-operative Organisation: Their impact on Traditional Fishermen of Kerala" Thesis, Centre for Development Studies, CDS Thiruvananthapuram.

Yusuf Muhammad (2013): "Innovation strategies and competitive forces to enter the European seafood market", Thesis, Faculty of Biosciences, Fisheries and Economics, Norwegian College of Fishery Science, Indonesia.

Working Papers

Ababouch, L., G. Gandini & J. Ryder (2005) Causes of Detentions and Rejections in International Fish Trade. FAO Fisheries Technical Paper 473. 110.

Aggarwal A (2002): "Anti-dumping Law and Practice: An Indian Perspective" Indian Council of Research on International Economic Relations (ICRIER) Working Paper No. 85

Aggarwal Aradhana(2003): " Patterns and Determinants of Antidumping: A World Wide Perspective" , Indian Council for Research on International Economic Relations, Working Paper No.113,New Delhi.

Cato, J.C. Economic values associated with seafood safety and implementation of seafood Hazard Analysis Critical Control Point (HACCP) programmes. FAO Fisheries Technical Paper. No. 381. Rome, FAO. 1998. 70p.

Chengyan Yue, John C. Beghin, and Helen H. Jensen (2005): "Tariff Equivalent of Technical Barriers to Trade with Imperfect Substitution and Trade Costs" Center for Agricultural and Rural Development, Iowa State University Ames, Iowa, Working Paper 05-WP 383.

Coughlin C Cletus and Wood E Georffrey(1989): "An Introduction to Non- Tariff Barriers to Trade "The Federal Reserve Bank of St Louis Review, January/ February 1989 , pp. 32-46

D. Rajasenan (2005), "Sanitary and Phytosanitary Agreement on Fish Trade: Challenges and Opportunities for Kerala's Exports", Working paper, (Kochi, CUSAT).

Frank a Martin and D. Smith (2009): "Trade and Fisheries: Key Issues for the World Trade Organisation", World Trade Organisation, Economic Research and Statistics Division, Staff Working Paper ERSD-2010-03 January 2010.

Frohberg Klaus, Grote Ulrike and Winter Etti(2006):" EU Food Safety Standards, Traceability and Other Regulations: A Growing Trade Barrier to Developing Countries' Expots",Paper Presentation, International Association of Agricultural Economist Conference, Gold Coast, Australia, August 12-18,2006.

Fugazza Marco and Christophe Maur Jean (2008): "Non- Tariff Barriers in Computable General Equilibrium Modeling", United Nations Conference on Trade and Development, Policy Issues in International Trade and Commodities, Study Series No. 38. New York and Geneva,

Grimwade Nigel," Anti-dumping Policy: An Overview of the Research", Centre for International Business Studies, research working Paper pp1-9

Henson Spencer, SaquibMohammed, Rajasenan D (2004): "Impact of Sanitary Measures on Exports of Fishery products from India: The Case of Kerala, Agriculture and Rural Development" Discussion Paper17,Cost of Compliance with SPS standards, World Bank,pp1.(www.worldbank.org)

Henson Spencer, Saqib Mohammed, Rajasenan D(2004): " Impact of Sanitary Measures on Exports of Fishery Products from India" The Case of Kerala, Cost of Compliance with SPS Standards, The International Bank for Reconstruction and Development / The World Bank Agriculture and Rural Development Discussion Paper 17.

Irwin A Douglas (2005): "The Rise of US Anti-dumping Activity in Historical Perspective "International Monetary Fund Working Paper, WP/05/31.

(IFAD, 2003): " International Fund for Agriculture Development, ACHIEVING THE MILLENNIUM DEVELOPMENT GOALS" by Enabling the Rural Poor to Overcome their poverty" Discussion Paper for the Twenty-Fifth Anniversary Session of IFAD's Governing Council(1978-2003), February 2003

John Kurien (1978): "Towards an Understanding of the Fish Economy of Kerala State" Working Paper No.68 Centre for Development Studies, pp61-72.

J Michael Finger (1991): "The Origins and Evolution of Antidumping Regulation", Policy Research and External Affairs, Country Economics Department, The World Bank, Working Papers, WPS 783.

KamatManasvi and KamatManoj (2007), "Implications of the WTO on Indian Marine Industry, Issues and Policy Perspectives" MPRA Paper No.6151,2007.

Koczan Zsoka and Plekhanov Alexander (2013): "How important are non-tariff barriers? Complementarity of infrastructure and institutions of trading partners" European Bank for Reconstruction and Development, Working Paper No. 159, London

Kumar C Nalin (2012): "Asian India FTA and Commodity Trade: Prioritzing invisible infrastructure Indian Institute of Plantation Management", Jnana Bharathi Campus, Bangalore.

Kulkarni Parashar (2005): "The Marine Seafood Export Supply Chain in India", Current State and Influence of Import Requirements, Consumer Unity and Trust Society (CUTS), The International Institute for Sustainable Development, Canada

Kurien John (2003): The Blessing of the Commons: Small-Scale Fisheries, Community Property Rights, and Coastal Natural Assets, CDS/ SIFFS Publication, Trivandrum, Working Paper 349.

Lamb E John,Julian A Velez, Barclay W Robert(2005): " The Challenge of Compliance with SPS and Other Standards Associated with the Export of Shrimp and Selected Fresh Produce Items to the United States Market", Agricultural and Rural Development Discussion Paper, The World Bank, Washington.

Low, P. and Yeats, A. J. (1994): "Non-Tariff Barriers and Developing Countries: Has the Uruguay Round Leveled the Playing Field?" Washington, D.C., World Bank, Policy Research Working Paper 1353

Levy I Philip (2003): "Non-Tariff Barriers as a Test of Political Economy Theories" Economic Growth Center, Yale University, Center Discussion Paper No.852, February 2003.

Mehta Rajesh (2005): " Non-Tariff Barriers Affecting India's Exports", Research and Information System for the Non-Aligned and Other Developing Countries (RIS) ,Discussion Papers, New Delhi.

Nandana Baruah (2005): "Anti Dumping As a measure of Contingent Protection: An Analysis of Indian Experience" Working Paper No.377, Centre for Development Studies.

Parvathy P and D Rajasenan (2012),"From Market Concentration to Market Diversification: WTO and the Marine Products Exports from Kerala- Using ARIMA Method" ISSN 2224-607X (Paper) ISSN 2225-0565(Online). Vol 2, No.6.2012.

Prusa, ThomasJ and Skeath Susan (2001): "The Economic And Strategic Motives for Antidumping Feelings" NBER Working Paper No. 8424.

Rajan B J(2002) " Labour Mobility in the Small-Scale Fisheries Sector of Kerala" Kerala Research Programme on Local Level Development, Centre for Development Studies, Discussion Paper No.44,Thiruvananthapuram.

Santos-Paulino U. Amelia (2012): "Trade, Income Distribution and Poverty in Developing Countries" United Nations Conference on Trade and Development, Discussion Papers, No. 207, July 2012.

Satish Y Deodhar(2003): " Motivation For and Cost of HACCP in Indian Food Processing Industry "Indian Institute of Management, Ahmadabad, Working Paper No. 2003-05-03

Saqib Mohammed and Taneja Nisha (2005): "Non-Tariff Barriers and India's Exports: The Case of Asean and Sri Lanka" Working Paper No. 165, Indian Council for Research on International Economic Relations, New Delhi.

Staiger W Robert. Stanford Wisconsin and NBER (2012) Non-Tariff Measures and the WTO, Economic Research and Statistics Division, World Trade Organization, Staff Working Paper ERSD-2012-01 dated, January 2012

Sumaila, U Rashid, Chrisatophe Bellmann and Alice Tipping (2014): Fishing for the Future: Trends and Issues in Global Fisheries Trade, E15 Initiative, Geneva: International Centre for Trade and Sustainable development (ICTSD) and World Economic Forum 2014. www.e15iniitiative.org

Van Tongeren, F., J. Beghin and S. Marette (2009), "A Cost-Benefit Framework for the Assessment of Non-Tariff Measures in Agro-Food Trade", OECD Food, Agriculture and Fisheries Working Papers, No. 21, OECD.

Willems Sabine, Roth Eva, Roekel van Jan (2005) :Changing European Public and Private Food Safety and Quality requirements Challenges for Developing Country Fresh Produce and Fish Exporters, European Union Buyers Survey Agriculture and Rural Development Discussion Paper, The World Bank.

Reports and Other Publications

Annon(2005): 'Poor logistics costing seafood industry Rs.6000 crores annually' Business Line, The Hindu December 15th Cochin Edition, Cochin.

Bacchetta Marc and Beverelli Cosimo (2012): 'Trade and public policies: A closer look at non-tariff measures in the 21st century' Economic Research and Statistics Division, World Trade Organization, October 01, 2012.

Bostock T, Greenhalgh P and Kleih.U (2004): 'Policy Research - Implications of Liberalization of Fish Trade for Developing Countries: Synthesis Report', Chatham, Natural Resources Institute, University of Greenwich, UK.

Brown, Drusilla K., Deardorff, Alan V. and Robert M. Stern (2002): Computational Analysis of Multilateral Trade Liberalization in the Uruguay Round and Doha Development Round. Discussion Paper No. 489. School of Public Policy. The University of Michigan. December 8, 2002.

CSO (2012): 'Quick Estimates of National Income. Consumption Expenditure, Saving and Capital Formation 2010-11', Central Statistics Office (CSO), Ministry of Statistics and Programme Implementation, Govt. of India, New Delhi, 31st January.

CMFRI (2010): "Marine Census" Kochi.

Degaldo L Christopher,WadaNikolas,Rosegrant W Mark, Meijer Siet and Ahmed Mahfuzuddin(2003), 'Outlook For fish to 2020 Meeting Global Demand" IFPRI,International Policy Research Institute, Food Policy Report Vision 2020'. World Fish Centre,Washington, D.C., U.S.A.

Economic Review (Various Issues): 'State Planning Board', Government of Kerala.

Emmerson, W.(2006): 'The Dispute Settlement Understanding' Power Point presentation, Geneva, March, 2006.

Gordon, A., Pulis, A., Owusu-Adjei, E. (2011): "Smoked marine fish from Western Region, Ghana: a value chain assessment", World Fish Center. USAID Integrated Coastal and Fisheries Governance Initiative for the Western Region, Ghana. 46pp.

District Handbook (2012): *"District Handbook of Ernakulatm District"*, Bureau of Economics and statistics, Government of Kerala.

Handbook on Fisheries Statistics (2014): Department of Animal Husbandry, Dairying and Fisheries, Ministry of Agriculture, Government of India.

Handbook on Fisheries and Aquaculture (2010): "Three phases of development of Indian marine fisheries" ICAR, New Delhi

ICAR (2013): Handbook of Fisheries and Aquaculture" Indian Council of Agricultural Research, New Delhi.

Harikumar G. & Rajendran G (2007): An Over View of Kerala Fisheries – with Particular Emphasis on Aquaculture, IFP Souvenir 2007.

Fergusson F Ian (2011): World Trade Organisation Negotiations: The Doha Development Agenda, Specialist in International Trade and Finance, Congressional Research Service, December 12,2011, 7-5700 www.crs.gov RL32060

Fugazza Marco (2013): "The Economics Behind Non-Tariff Measures: Theoretical Insights and Empirical Evidence "Policy Issues in International Trade and Commodities, Study Series No.57, UNCTAD, Geneva.

FAO(2008),FAO Fisheries – The State of World Fisheries and Aquaculture,2008PART 1:World review of fisheries and aquaculture, Fish consumption, p. 58-65

FAO Fishstat (2009): Fishstat Plus – Universal Software for fishery statistical time series

FAO (Various Issues): The State of Food Insecurity in the World- Economic crises – Impacts and lessons learned, The United Nations.

FAO (2010): The State of world fisheries and aquaculture, Rome, pp.24

FAO(2013),"The State of Food Insecurity in the World 2013 The Multiple dimension of food security" Food and Agriculture Organization of the United Nations, Rome,2013.

FAO (2012): "World Review of Fisheries and Aquaculture, Status and Trends Part" The State of World Fisheries and Aquaculture 2012.

FAO Statistical Yearbook (2013): 'World Food and Agriculture' Food and Agriculture Organization of the United Nations, Rome 2013.

Government of India (2011), Faster, Sustainable and More Inclusive Growth: An Approach to the 12th Five Year Plan (2012- 2017), Planning Commission, New Delhi: Government of India

Government of India (2011b). Agricultural Statistics at a Glance 2011, Directorate of Economics & Statistics, Department of Agriculture & Cooperation, Ministry of Agriculture, Govt. of India, New Delhi, October 2011.

Henson, S. and Loader, R. (2001). Barriers to Agricultural Exports from Developing Countries: The Role of Sanitary and Phytosanitary Requirements. World Development, 29(1): 85-102.

Hussan F,Jeeva Charles and Prathap Sangeetha K(2012),'Economics of cost of compliance with HACCP in seafood export units and its limitations for applicability in domestic markets", Central Institute of Fisheries Technology, Cochin, India.

Josupeit, H. (2005) WTO and Fisheries. Globe Fish, October, 2005.

Kachru R. P (2006): "Agro-Processing Industries in India—Growth, Status and Prospects" Asstt. Director General (Process Engineering), Indian Council of Agricultural Research, New Delhi

Khatun Fahmida(2004) " A Case Study for Bangladesh Fish Trade Liberalisation in Bangladesh: Implications of SPS Measures and Eco- Labelling for the Export Oriented Shrimp Sector, Policy Research- Implications of Liberalisation of Fish Trade for Developing Countries, Food and Agriculture (FAO) of the United Nations, Rome, July 2004, Project PR 26109.

Kommerskollegium /National Board of Trade, Sweden: Report (2005-02-10) The Use of Antidumping in Brazil, China, India and South Africa – Rules, Trends and Causes

Kurien (1974): "Towards an Understanding of Kerala's Fish Economy", Working paper, CDS Working Paper 68, Trivandrum

Lem Audun,Bjorndal Trond,Lappo Alena (2014) "Economic Analysis of Supply and Demand for Food Up to 2030- Special Focus on Fish and Fishery Products "Food and Agriculture Organisation of the United Nations, Fisheries and Aquaculture Circular No.1089 Rome.

Ministry of Commerce & Industry (2015): F. No. 1(11)/2014-EPL, Department of Commerce, Economic Division, Government of India, New Delhi, Dated 15th January, 2015

MPEDA (2013), Marine Products Export Development Authority, Kochi.

NFDB (2009), Guidelines for domestic marketing, NFDB, India.

Nair, Balakrishnan, B, 2002: "Report of the Expert Committee for Fisheries Management Studies, Kerala", (Trivandrum, Directorate of Fisheries).

NAAS (2006): WTO and Indian Agriculture: Implications for Policy and R&D, Policy Paper No. 38, National Academy of Agricultural Sciences, New Delhi. pp 12.

NFPDB (2012): "Draft Detail Project Report", National Fish Producing Development Board, Ministry of Food Processing Industries, New Delhi.

Ministry of Commerce & Industry (2015): F. No. 1(11)/2014-EPL, Department of Commerce, Economic Division, Government of India, New Delhi, Dated 15th January, 2015

McCorriston Steve and MacLaren Donald (2012),"State Trading Enterprises as Non-Tariff Measures: Theory, Evidence and Future Research Directions", The Global Trade Analysis Conference, Geneva, 27th, 29th June, 2012.

Mehta R, Saqib M, George J (2002): "Addressing Sanitary And Phytosanitary Agreement:A Case Study of Select Processed Food Products in India" Research and Information System for the Non-Aligned and Other Developing Countries – RIS-DP # 39/2002.

Morey Phillip (2008): "Seafood Market Supply Chain – South East Sulawesi" Fisheries and Aquaculture Department of Primary Industries Victoria, Morelink Asia Pacific, 18th June 2008.

Ministry of Commerce & Industry (2015): F. No. 1(11)/2014-EPL, Department of Commerce, Economic Division, Government of India, New Delhi, Dated 15th January, 2015

Shyam, S Salim (2012) *Indian Seafood industry and post WTO- A Policy Outlook.* In: World Trade Agreement and Indian Fisheries Paradigms: A Policy Outlook, 17-26 September 2012, Kochi.

Silas E G(1977): "Indian Fisheries 1947-977" issued on the occasion of the Fifth session of the Indian Ocean Fishery commission held at Cochin from19th to 26th December, 1977.

Szulecka Olga (2006): "The Difficulties in traceability systems implementation in the Polish Fish Industry" Sea fisheries Institute in Gdynia Peter Workshop Consolidating and Exploiting EU Research on Food and Feed Traceability, CSL, York, UK

MPEDA (2013): .Marine Products Export Development Authority, Kochi.

Roheim A Cathy (2004): "Trade Liberalization in Fish Products: Impacts on Sustainability of International Markets and Fish Resources." In: Global Agricultural Trade and Developing Countries, A. Aksoy and J. Beghin, eds. The World Bank, Washington, DC: 275-295.

Salagarma Venkatesh and Thaddeus Koriya(2006), "Sustainability Impact Assessment of Proposed WTO Negotiations: The Fisheries Sector" ICM Integrated Coastal Management, Andhra Pradesh ,India.

SEAI (2010): Seafood Export Association of India, 40th Annual Report, 2009-2010.

Sumaila, U Rashid, Chrisatophe Bellmann and Alice Tipping (2014): Fishing for the Future: Trends and Issues in Global Fisheries Trade, E15 Initiative, Geneva: International Centre for Trade and Sustainable development (ICTSD) and World Economic Forum 2014. www.e15iniitiative.org

The Hindu -Business Line (2001): "Move to build brand equity for seafood exports", Visakhapatnam, Feb. 8

Trondsen Torbjorn (2003): "Criteria and methodology to improve the effects of international trade on food security in fish-exporting and fish-importing developing countries" Expert Consultation on International Fish Trade and Food Security, Corporate Document Repository, FAO.

Unnevehr, L. J. (1999). Food safety issues and fresh food product exports from LDCs. Paper presented at the conference: Agro-Industrialization, Globalisation and Development, Nashville, August.

UN (2007): World Population Prospects: The 2006 Revision, Population Division of the Department of Economic and Social Affairs of the United Nations Secretariat, table 1.1. pp.1

UNCTAD (2012): 'United Nations conference on Trade and Development, Classification of Non-tariff measures' February 2012 version.

FAO (1995): 'Code of Conduct f\or Responsible fisheries', Food and Agriculture Organisation,

UNCTAD (2013): "Non-Tariff Measures to Trade: Economics and Policy Issues for Developing Countries" Developing Countries in International Trade Studies, United Nations Conference on Trade and Development, New York and Geneva.

Wilson, J.S, Sewadeh, M and Otsuki, T. (2001): 'Dirty Exports and Environmental Regulations: Do Standards Matter? World Bank.

WTO, (1995), The results of the Uruguay Round of Multilateral Trade Negotiations. The Legal Texts, Geneva, WTO.

The World Trade Organization (2005): "Legal, Economic and Political Analysis" Geneva, pp 231-370

WTO (2014): World Trade Organisation, *Annual Report* 2014, Geneva

Willems Sabine, Roth Eva, Roekel van Jan (2005) :Changing European Public and Private Food Safety and Quality requirements Challenges for Developing Country Fresh Produce and Fish Exporters, European Union Buyers Survey Agriculture and Rural Development Discussion Paper, The World Bank.

X Li and Sayed Saghaian (2011): "Impact of HACCP on U.S. Seafood Exports" Department of Agricultural Economics, Lexington.

U.S. International Trade Commission (2003): The Impact of Trade Agreements: Effect of the Tokyo Round, U.S.-Israel FTA, U.S.-Canada FTA, NAFTA, and the Uruguay Round on the U.S. Economy. Publication 3621. August 2003.

Zacharia P.U and Najmudeen T.M (2012): "Marine Finfish Resources of India: Distribution, Commercial Exploitation, Utilization Pattern and Trade" Central Marine Fisheries Research Institute, Kochi

Zheng Yuqing, Muth K Mary, Brophy Jenna(2013) "The Impact of Third-Party Certifications on Food Exports to the United States" Agricultural & Applied Economics Association's 2013 AAEA & CAES Joint Annual Meeting, Washington, DC, August 4-6, 2013.

Internet Sources

Agarwal Baya(2014): "India gets its first MSC certified fishery in Kerala" Sustainable Business http://www.wwfindia.org/?12341/India-gets-its-first-MSC-certified-fishery-in-Kerala access on November 5,2014.

Allain Marc (2007),"Trading Away Our Oceans- report why trade liberalization of fisheries must be abandoned" Greenpeace International, Netherlands.http://www.wto.org/english/forums_e/ngo_e/posp66_greenpeace_ocean_e.pdf accessed on 15/11/2013

Bhattarcharya B(2010):"The Indian Shrimp Industry Organizes to Fight the Threat of Anti-dumping Action", Managing the Challenges of WTO Participation: Case Study 17. http://www.wto.org/english/res_e/booksp_e/casestudies_e/case17_e.htm, accessed on October, 12 2010

Biswas Rajiv, Asia Pacific Chief Economist, forecast for Asia Pacific Region at IHS, a global information and analytics firm 2/8/2014. http://dw.de/p/1Cob4 accessed on October 24, 2014

Business Standards http: //www.business-standard.com/article/markets/india-is-largest-shrimp-exporter-to-us-114022700365_1.html accessed as on October 9t 2014Canadian Food Inspection Agency http://www.inspection.gc.ca/food/fish-and-seafood eng// 1299799645255/1299799784160 accessed on March 4, 2012

Cattermoul, B.,Brown, D. and Poulain, F. (2014): "Fisheries and aquaculture emergency response guidance" Rome, FAO. 167 pp.http://www.fao.org/3/a-i3432e.pdf accessed on 5th January, 2015.

Chen Shin-Chang, Chuang Ta Ching, Hu Sing-Hwa, and Nan Fan-Hua (2006): "Competitiveness and Supply Chain Management Study on Taiwan Grouper Industry" Competitiveness and SCM on Taiwan Grouper Industry, SCMIS.

http://www.agnet.org/htmlarea_file/activities/20110719101541/9.pdf accessed on 27th December 2010.

Cleins C. Coughlin and Geoffrey E. Wood," An Introduction to Non-Tariff Barriers to Trade" FEDERAL RESERVE BANK OF ST. LOUIS,1989.Pp 32-45. http://research.stlouisfed.org/publications/review/89/01/Trade_Jan_Feb1989.pdf accessed on December 12th,2012.

COGSI (2013): Coalition of Gulf Shrimp Industries Fact Sheet on Shrimp Subsidy Trade Cases.http://www.gulfshrimpcoalition.com/pdfs/RevisedCOGSIFactSheetFormattedAugust1.pdf accessed on 10th January 2013.

FAO (2012): The State of World Fisheries and Aquacutlure 2012 -Overview http://www.fao.org/fishery/publications/2012/en accessed on September 25th,2012 FAO (2014) World review of fisheries and aquaculture, Food and Agricultural Organization, Romehttp://www.fao.org/docrep/008/y5924e/y5924e06.htm accessed on September 20, 2014.

Food Safety of Chinahttp://en.wikipedia.org/wiki/Food_safety_in_China accessed on 18/10/2010

Gregory H Watson(2005): "Feigenbaum's Enduring Influence" Gurus of Quality

http://www.gregoryhwatson.eu/images/8-QP_Watson_-_November2005 Feigenbaum_'s_Enduring_Influence.pdf accessed on 10th December 2013.

GSP(2004):' The European Union's Generalised System of Preferences', European Commission ,Directorate-General for Trade.http://trade.ec.europa.eu/doclib/docs/2004/march/tradoc_116448.pdf accessed on July 2011.

Imani Development, Inventory of Regional Non-Barriers: Synthesis Report (2007); UNCTAD, Development and Globalisation: Facts and Figures (2008)

http://www.intracen.org/BB-2010-07-16-Address-non-tariff-measures-encountered-in-export-markets---Part-1/

International Centre for Trade and Sustainable Development. http://www.ictsd.org/bridges-news/bridges/news/wto-raises-trade-growth-forecasts-for-2014-2015 accessed on October 22, 2014

International Trade Administration (2013): Fact Sheet, U.S. Department of Commerce http://enforcement.trade.gov/download/factsheets/Factsheet-Frozen-Warmwater-Shrimp-CVD-Final-13AUG13.pdf accessed on 2nd January 2015.

Jawara, F.,Kwa,A.(2004) Behind the Scenes at the WTO, updated edition, London, Zed Books. http://www.michiganlawreview.org/articles/jawara-kwa-behind-the-scenes-at-the-wto-the-real-world-of-trade-negotiations-the-lessons-of-cancun acessed on 15/11/2013

K. Narayanan and Lalithambal Natarajan, "Globalisation, Trade Flows and Anti Dumping: Recent Indian Experience" , Department of Humanities & Social Sciences, I.I.T. Bombay, Powai. http://www.policyinnovations.org/ideas/policy_library/data/01185/_res/id=sa_File1/ accessed on October 23rd, 2014

Krishnakumar P K (2013): "Shrimps form 70 Percent of seafood export basket on globalshortage"EconomicTimesBureau http://articles.economictimes.indiatimes.com/2013-11-13/news/44031102_1_vannamei-seafood-export-basket-export-earnings accessed on 13th January 2014.

Millennium Goals Report (2014): http://www.un.org/millenniumgoals/2014%20MDG%20report/MDG%202014%20English%20web.pdf Accessed on 13 January 2015.

Ministry of State of Commerce & Industry, Dr. D. Purandeswari in a written reply in the Lok Sabha March 4th, 2013http://pib.nic.in/newsite/erelease.aspx?relid=92958 accessed on 31 March,2013.

Nataraj Geethanjali,"G20 and India" Observer Research Foundation, Delhi,2013. http://orfonline.org/cms/sites/orfonline/modules/analysis/AnalysisDetail.html?cmaid=55806&mmacmaid=55807 accessed on January 5th 2014.

): The EU Rapid Alert System for Food and Feed. http://ec.europa.eu/food/safety/rasff/index_en.htm accessed on 2January 2013.OECD–FAO (2013): ' The Agricultural Outlook, 2013-2022" http://www.oecd.org/site/oecd-faoagriculturaloutlook/highlights-2013-EN.pdf accessed on Novemeber 3rd 2014

Planning Commission (2001): "Report of the Working Group of Fisheries for the Tenth Five Year Plan Government of India" TFYP WORKING GROUP Sr. No. 16/2001.http://planningcommission.nic.in/aboutus/committee/wrkgrp/fishery.pdf accessed on January 13th ,2013.

RASFF - Food and Feed Safety Alerts(2013

Rendleman Matthew Charles and Vasinvarthana Yada (2013): "Is the Bioterrorism Act a Barrier to Developing Country Seafood Exporters" Southern Illinois University, Carbondale. file:///C:/Users/USER/Downloads/SSRN-id2269794. Accessed on November 12th 2013.

Report of the APO Study Meeting on Quality Enhancement in Small and Medium Food Processing Enterprises through HACCP (Hazard Analysis and Critical Control Point) held in India, 26 February–4 March 2002 (02-AG-GE-STM-01) www.apo-tokyo.org/publications/wp-content/uploads/.../agr-14-haccp.pdf accessed on September 20,2014.

Ruggero Urbani (2014): FAO Consultant, Fish Inspector, Quality Assurance Expert Ruggero.Urbani@fao.orghttp://ec.europa.eu/fisheries/documentation/publications/eu-new-fish-and-aquaculture-consumer-labels-pocket-guide_en.pdf accessed on 10th December 2013.

Seafood Exporters Association of India: http://seai.in/history/ accessed in March, 11th 2005.

Sehgal K.L (2012): "Cold Water Fish and Fisheries in the Western Ghats", NRC on Coldwater Fisheries (ICAR), Haldwani 263 139, Distt. Nainital (U.P.), India Sehgal K. L.."Coldwater fish and fisheries in the Western Ghats, India". FAO. http://www.fao.org/docrep/003/x2614e/x2614e06.htm Retrieved 24 August2012.

Singh Jeet, Yadav Preeti (2009): " Impact of Global Recession on Indian Economy with special reference to India's Exports" Management Insight , Vol. V, No. 2; December, 2009 http://www.smsvaranasi.com/insight/impact_of_global_recession_on_indian_economy_with_special_reference_to_indias_exports.pdf accessed on March 2012.

SOFIA (2010): The state of World Fisheries and aquaculture 2010, SEAFISH, Headline news, FAO, issued January 2011 http://www.fao.org/docrep/013/i1820e/i1820e00.htmaccessed on December 12th ,2012.

Southern Shrimp Alliance, Eighth Administrative Reviews of the antidumping Duty Orders on Shrimp, http://www.shrimpalliance.com/commerce-announces-preliminary-results-in-eighth-administrative-reviews-of-the-antidumping-duty-orders-on-shrimp/ accessed on October 2nd, 2014.

http://www.shrimpalliance.com/news-alert-u-s-international-trade-commission-releases-final-determination-in-countervailing-duty-investigations/ accessed on October 10th, 2014.

T. B. Simi (2008): "India, Thailand and US on Anti-dumping Measures relating to Shrimp Another case calling for clarity in the WTO rules", CUTS Centre for International Trade, Economics & Environment No.2/2008 http://www.cuts-citee.org/pdf/TLB08-02.pdf accessed May 3, 2012

Thomas W. Hertel and Roman Keeney, "What is at Stake: The Relative Importance of Import Barriers, Export Subsidies and Domestic Support," in Anderson and Martin, eds., Agricultural Trade Reform in the Doha Agenda (Washington: World Bank, 2005); and Kym Anderson, Will Martin, and Dominique van der Mensbrugge, "Doha Merchandise Trade Reform: What's At Stake for Developing Countries," July 2005, available at http://www.worldbank.org/trade/wto. accessed on 5th January 2013.

UN-DESA.(2009):World population to exceed 9 billion by 2050",http://www.un.org/esa/population/publications/wpp2008/pressrelease.pdf. accessed on 5th January 2013.

United Nations (2012): "The future we want". www.uncsd 2012.org. accessed on 5th January 2013.

United Nations (2014): "Millennium Development Goals and post -2015 Development Agendas", www.un.org/en/ecosoc/anout /mdg.shtml accessed on 5th January 2013.

UNCTAD (2013): "Classification of Non- Tariff Measures" United Nations Conference on Trade and Development, New York and Geneva, February 2012 Version. accessed on 5th January 2013.

http://unctad.org/en/PublicationsLibrary/ditctab20122_en.pdf accessed on 5th January 2013.

U.S. Food and Drug Administration(2015): " President's FY 2016 Budget Request: Key Investments for Implementing the FDA Food Safety Modernization Act (FSMA)"

http://www.fda.gov/Food/GuidanceRegulation/FSMA/ucm432576.htm?source=govdelivery&utm_medium=email&utm_source=govdelivery accessed on February 2nd, 2015.

USDA United States Department of Agriculture Economic Research Service.

http://www.ers.usda.gov/data-products/aquaculture-data.aspx Accessed on April 17, 2015.

US upholds Anti-dumping duties against Thai and Indian shrimphttp://www.ictsd.org/bridges-news/biores/news/us-upholds-anti-dumping-duties-against-thai-and-indian-shrimp accessed on 23rd December 2012.

Vladimir Popov (2007)**: Food Safety in Russia**. International Seminar. "Emerging Food Safety Risk: How can we know?" 27 November 2007, Bangkok, Thailand. file:///C:/Users/USER/Downloads/PopovFoodSafetyinRussia%20(1).pdf accessed on March 2, 2010

Valdimarsson Grimur(2011):"Challenges for the Global Seafood Industry "World trade in seafood: key trends and issues Food and Agriculture Organization of the United Nations, Romeftp://ftp.fao.org/docrep/fao/011/a1293e/a1293e01.pdf accessed on September20,2014.

WTO (2001): Ministerial Declaration, Doha. WT/MIN (01)/DEC/1.

(http://www.wto.org).WTO Report 2013, Accessed on 3 March, 2009.

World Trade Report (2012): World Trade Organization.http://www.wto.org/english/res_e/booksp_e/anrep_e/world_trade_report12_e.pdf accessed on Novemeber 12th,2014.

WTO(2014):World Trade Organization Annual Report https://www.wto.org/english/res_e/booksp_e/anrep_e/anrep14_e.pdf accessed on 1st March 2015.

WTO (2014): Understanding WTO, World Trade Organisation http://www.wto.org/english/thewto_e/whatis_e/tif_e/fact4_e.htm accessed on October 22, 2014

WTO Annual Report 2014. http://www.wto.org/english/res_e/booksp_e/anrep_e/anrep14_e.pdf accessed on October 22, 2014.

APPENDICES

APPENDIX 1

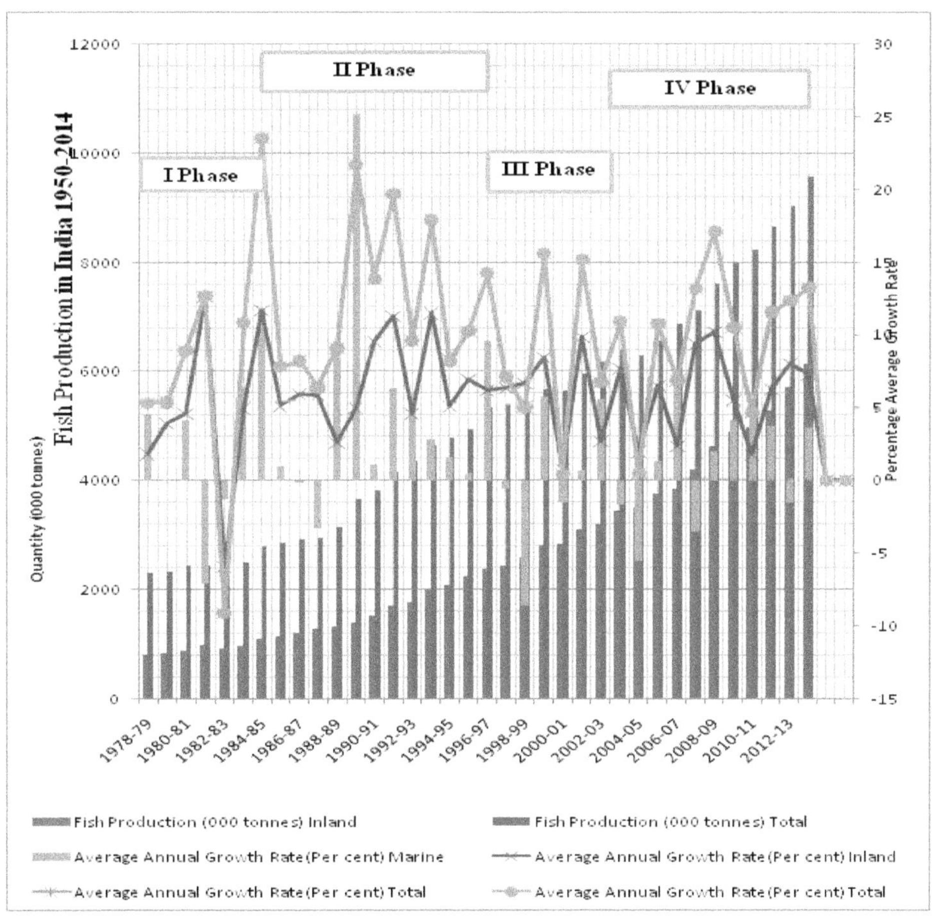

Source: State Governments/ Union Territory Administrations(P): Provisional

i. Central Marine Fisheries Research Institute, Kochi for the period up to 1970-71

ii. State Governments/Union Territory Administrations since 1971-72

APPENDIX 2

FISH SEED PRODUCTION OVER DIFFERENT PLAN PERIODS IN INDIA		
PLAN PERIOD	YEAR	PRODUCTION (Million Fry)
End of 4th Plan	1973-74	409
End of 5th Plan	1978-79	912
End of 6th Plan	1984-85	9639
7th Plan	1985-86	6322
	1986-87	7601
	1987-88	8608
	1988-89	9325
	1989-90	9691
Annual Plan	1990-91	10332
	1991-92	12203
8th Plan	1992-93	12500
	1993-94	14239
	1994-95	14544
	1995-96	15007
	1996-97	15852
9th Plan	1997-98	15904
	1998-99	15156
	1999-00	16589
	2000-01	15608
	2001-02	15758
10th Plan	2002-03	16333
	2003-04	19231
	2004-05	20791
	2005-06	21988
	2006-07	23648
11th Plan	2007-08	24144
	2008-09	32177
	2009-10	29313
	2000-11	34111
	2011-12	36566
12th Plan	2012-13	34922
	2013-14	41450
Source : Director/Commissioner of Fisheries State Govts./UT Administrations (P): Provisional		

APPENDIX 3: MAJOR EVENTS OF WTO

Year	Events
1994	Marrakesh Agreement Establishing in the WTO is signed
1995	WTO is born on 1st January
	Renato Ruggiero (Italy) takes office as WTO Director General
1996	First Ministerial Conference Takes place in Singapore
1997	70 WTO members reach a multilateral Agreement to open their financial services
1998	Second Ministerial Conference taken place in Geneva
1999	Mike Moore(New Zeland) become WTO Director General,Third Ministerial Conference takes place in Seattle US
2000	Negotiation begins on Services ,Negotiations begins on Agriculture.
2001	Fourth Ministerial Conference in Doha Quatar, Doha Development Agenda is launched
2002	Supachai Panitchpakdi (Thailand) is elected WTO Director General
2003	Fifth Ministerial Conference Takes place in Cancun Mexico
2004	Ministerial Discussion on the Doha Round takes place in Geneva
2005	Pascal Lamy (France) becomes WTO Director General,Sixth Ministerial Conference takes place in Hong Kong. China Aid for trade initiation is launched. Hong Kong declaration is approved. Sixth Ministerial Conference takes place in Hong Kong. China Aid for trade initiation is launched. Hong Kong declaration is approved
2006	Ministerial Discussion on the Doha Round takes place in Geneva, First Ministerial Conference Takes place in Singapore
2007	Vietnam becomes the WTO'S150th member, First Global Review of Aid for trade takes place in Geneva
2008	Cabo Verde joins the WTO, Ministerial Discussion on the Doha Round takes place in Geneva
2009	DG Pascal Lamy reappointed for seond term of o four years, Second Global Review of Aid for Trade takes place in Geneva and Seventh Ministerial Conference takes place in Geneva
2010	New "Chairs Programme" launched to support developing country universities, Second WTO Open Day in Geneva
2011	Eighth Ministerial Conference takes place in Geneva, Membership agreement where made for Russia, Samoa, and Montenegro, dependent on the ratification of those countries
2012	Montenegro, Samoa , Russia and Vanuatu join the WTO
2013	Laos joins the WTO and Tajikistan becomes the 159th member of the WTO. Roberto Azevedo (Brazil) takes office as WTO Director General. Ninth Ministerial Conference takes place in Bali, Indonesia, Bali Package adopted.
2014	Yemen Become 160th WTO Member. General Council approves Seychelles, first non-least-developed African country to join the Organization' WTO membership, only ratification left and hopefully, become the 161st WTO member upon accession. Revised WTO Agreement on Government Procurement to come into force on 6 April 2014
2015	WTO is commemorating its 20th anniversary.
Source: WTO Report	

APPENDIX 4

Top Ten Countries in Landings in 2012

Country	Landings(000tonnes)	Percent of global Landings
China	13,870	17
Indonesia	5,420	7
US	5,131	6
Peru	4,804	6
Russian Federation	4,069	5
Japan	3,611	5
India	3,402	4
Chile	2,573	3
Vietnam	2,419	3
Myanmar	2,332	3

Source FAO 2014

Top Ten Countries in Landed Value in 2006

Country	2005 real value(million USD)	Percent of global Landed Value
China	13,293	12
Japan	8,354	8
US	8,248	8
Indonesia	6,202	6
Peru	4,645	4
Thailand	3,743	3
Chile	3,704	3
Korea Rep	3,167	3
India	3,004	3
Spain	2,506	2

Source: Sea around Us (Watson et al 2006 Sumaila et al, 2007 and Swartz et al)

APPENDIX 5

Nutritional Value of Fish
Fish can be divided into 3 groups: Oily Fish, White Fish, and Shellfish

OILY FISHES	WHITE FISHES	SHELLFISHES
Salmon	Cod	Crabs
Mackerel	Plaice	Mussels
Fresh Tuna	Whiting	Oysters
Trout	Haddock	Lobster
Sardines	Sole	Prawns
Herring	Hake	

Oily Fish

- Rich Source of Vitamins A, D, and E. Filled with essential Omega-3 fatty acids
- Essential for health brain, eye and nerve development in babies and children
- Beneficial for heart health, emerging evidence to suggest that eating fish reduces the risk of cancer and arthritis.

White Fish

- Improves blood pressure lowers cholesterol and reduces weight, according to a study conducted by researchers at the Nutrition & Obesity Unit of Hospital Ramon y Cajal in Madrid.

Shellfish

- Low in fat: Lobster and shrimp have less than 1 g of fat per serving and very little of that in saturated fat.
- Most shellfish (except for shrimp) are also very low in cholesterol. Shellfish can be a low-fat, Low- Cholesterol, Heart- healthy choice.
- Good Source of protein.
- Full of healthy vitamins A, B1, B2, B3 and D. Also high in minerals like iron, calcium and magnesium.
- One of the best dietary sources of zinc, a mineral necessary for keeping your immune system healthy and promoting the healing of wounds. The highest levels of zinc can be found in oysters.

APPENDIX 6

Some Popular Fish in India

Kerala

Chaala(Sardines)
Ayala(Mackerel)
Aavoli(Black Pomfret)
Neymeen(Seer Fish)
Choora(Tuna)
Kanirmeen(Black Pearl Spot)

Maharashtra & Mangalore

Bombil(Bombay Duck)
Teesriya(Clams)
Kekda(Crabs)
Kane(Lady Fish)
Bangda(Mackerel)
Raavas(Salmon)
Surmai(Seer Fish)

Bengal

Rui(Rohu)
Koi(Climbing Perch)
Pink- belly Indian
Butter Fish
Magur(Catfish)
Hilsa

APPENDIX 7

World fish market at a glance

	2012	2013 estim.	2014 estim.	Change: 2014 over 2013
		million tonnes		%
WORLD BALANCE				
Production	**157.8**	**162.8**	**165.9**	**1.9**
Capture fisheries	91.3	92.6	92.0	-0.6
Aquaculture	66.5	70.2	73.9	5.3
Trade value (exports USD billion)	**129.3**	**136.5**	**143.9**	**5.4**
Trade volume (live weight)	**58.1**	**58.8**	**59.5**	**1.2**
Total utilization	**157.8**	**162.8**	**165.9**	**1.9**
Food	136.0	141.0	144.6	2.6
Feed	16.3	16.8	16.6	-1.2
Other uses	5.4	5.0	4.7	-6.0
SUPPLY AND DEMAND INDICATORS				
Per caput food consumption:				
Food fish (kg/year)	19.2	19.7	20.0	1.4
From capture fisheries (kg/year)	9.8	9.9	9.8	-1.2
From aquaculture (kg/year)	9.4	9.8	10.2	4.1

Totals may not match due to rounding.

APPENDIX 8

Total and per capita fish consumption by continent in 2011*
140
120
100
80
60
40
20
0
130.1
18.9
89.8
21.6
16.2
22.0
9.9
9.7
7.5
21.8
5.7
9.7
0.9
25.4
World
Asia
Europe
Africa
North America
Latin America and the Caribbean
Oceania
Total (million tonnes)
Per capita (kg/year)

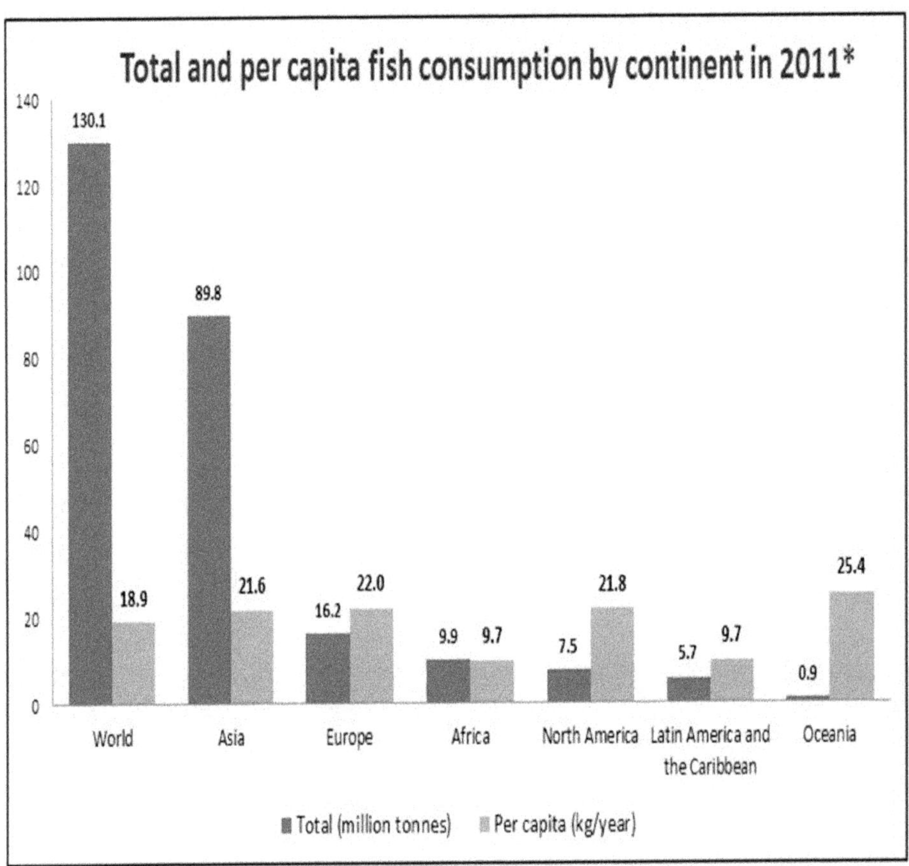

www.ingramcontent.com/pod-product-compliance
Ingram Content Group UK Ltd.
Pitfield, Milton Keynes, MK11 3LW, UK
UKHW021532300726
14060UKWH00011B/375

9 789387 0577